T0191965

Advances in Geographical and Environmental Sciences

Series Editor

R. B. Singh, University of Delhi, Delhi, India

Advances in Geographical and Environmental Sciences synthesizes series diagnostigation and prognostication of earth environment, incorporating challenging interactive areas within ecological envelope of geosphere, biosphere, hydrosphere, atmosphere and cryosphere. It deals with land use land cover change (LUCC), urbanization, energy flux, land-ocean fluxes, climate, food security, ecohydrology, biodiversity, natural hazards and disasters, human health and their mutual interaction and feedback mechanism in order to contribute towards sustainable future. The geosciences methods range from traditional field techniques and conventional data collection, use of remote sensing and geographical information system, computer aided technique to advance geostatistical and dynamic modeling.

The series integrate past, present and future of geospheric attributes incorporating biophysical and human dimensions in spatio-temporal perspectives. The geosciences, encompassing land-ocean-atmosphere interaction is considered as a vital component in the context of environmental issues, especially in observation and prediction of air and water pollution, global warming and urban heat islands. It is important to communicate the advances in geosciences to increase resilience of society through capacity building for mitigating the impact of natural hazards and disasters. Sustainability of human society depends strongly on the earth environment, and thus the development of geosciences is critical for a better understanding of our living environment, and its sustainable development.

Geoscience also has the responsibility to not confine itself to addressing current problems but it is also developing a framework to address future issues. In order to build a 'Future Earth Model' for understanding and predicting the functioning of the whole climatic system, collaboration of experts in the traditional earth disciplines as well as in ecology, information technology, instrumentation and complex system is essential, through initiatives from human geoscientists. Thus human geosceince is emerging as key policy science for contributing towards sustainability/survivality science together with future earth initiative.

Advances in Geographical and Environmental Sciences series publishes books that contain novel approaches in tackling issues of human geoscience in its broadest sense — books in the series should focus on true progress in a particular area or region. The series includes monographs and edited volumes without any limitations in the page numbers.

More information about this series at https://link.springer.com/bookseries/13113

Swapan Kumar Maity

Essential Graphical Techniques in Geography

 Springer

Swapan Kumar Maity
Department of Geography
Nayagram P.R.M. Government College
Jhargram, West Bengal, India

ISSN 2198-3542 ISSN 2198-3550 (electronic)
Advances in Geographical and Environmental Sciences
ISBN 978-981-16-6587-5 ISBN 978-981-16-6585-1 (eBook)
https://doi.org/10.1007/978-981-16-6585-1

This Springer imprint is published by the registered company Springer Nature Singapore Pte Ltd.
The registered company address is: 152 Beach Road, #21-01/04 Gateway East, Singapore 189721,
Singapore

Dedicated to my Parents

Preface

Geography is a scientific discipline that emphasizes how and why different geographic features vary from one place to another and how spatial patterns of these features change with time. Geographers always concentrate on the explanation of how physical and cultural features are distributed on the earth surface and what kinds of factors and processes are responsible for their spatial and temporal variations. Geographical data need appropriate, systematic and logical presentation for a better understanding of their cartographic characteristics. Suitable, accurate and lucid demonstration and visualization of geographical data become helpful for their correct analysis, explanation and realization. Therefore, various types of primary and secondary data are used voluminously to explain and analyze the spatial distributions and variations of different geographical events and phenomena.

Graphs, diagrams and maps are three unique and distinctive techniques of visualization of geographical data. In narrow sense, graphical representation means the depiction of data using various types of graphs but in a wider sense, all types of graphs, diagrams and mapping techniques are included in graphical methods of portraying the data. Graphical representation of various kinds of geographical data is very simple, attractive and easily understandable not only to the geographers or efficient academicians but also to the common literate people. It is the key for geographers and researchers to recognize the nature of data, the pattern of spatial and temporal variations and their relationships and the formulation of principles to accurately understand and analyze features on or near the earth's surface. These modes of representation also enable the development of spatial understanding and the capacity for technical and logical decision making.

In this book, attempts have been made to analyze and explain different kinds of graphs, diagrams and mapping techniques, which are extensively used for the visual representation of various types of geographical data. The book has broadly been divided into four main chapters. Chapter 1 emphasizes the discussion of the concept and types of geographical data, major differences between them, sources of each type of data, methods of their collection, classification and processing of the collected data with special emphasis on frequency distribution table, methods and appropriateness of representation of data and advantages and disadvantages of using these methods.

It includes the discussion of the concept of attribute and variable, types of variables and differences between them. It also explains different types of measurement scales used in geographical analysis.

Chapter 2 includes the detailed classification of all types of graphs and types of co-ordinate systems with illustrations as an essential basis of construction of graphs. Different types of Bi-axial (Arithmetic and Logarithmic graph, Climograph etc.), Tri-axial (Ternary graph), Multi-axial (Spider graph, Polar graph etc.) and special graphs (Water budget graph, Hydrograph, Rating curve, Lorenz curve, Rank-size graph, Hypsometric curve etc.) have been discussed with suitable examples in terms of their appropriate data structure, necessary numerical calculations, methods of construction, proper illustrations and advantages and disadvantages of their use. Concept of arithmetic and logarithmic graphs has been explained precisely with pertinent examples and illustrations. Different types of frequency distribution graphs have been explained with suitable data, necessary mathematical and statistical computations and proper illustrations.

Chapter 3 focuses on the detailed discussion of various types of diagrams classified on a different basis. All types of one-dimensional (Bar, Pyramid etc.), two-dimensional (Triangular, Square, Circular etc.), three-dimensional (Cube, Sphere etc.) and other diagrams (Pictograms and Kite diagram) have been discussed with suitable examples in terms of their appropriate data structure, necessary numerical (geometrical) calculations, methods of construction, appropriate illustrations and advantages and disadvantages of their use.

Chapter 4 explains the basic Cartographic terminologies like Geodesy, Geoid, Spheroid, Datum, Geographic co-ordinate system, Surveying and levelling, Traversing, Bearing, Magnetic declination and inclination etc in a lucid manner with suitable illustrations. It includes the detailed classification and discussion of all types of maps based on their scale and purposes (contents) of preparing the map with special emphasis on Indian Topographical Sheets. All pictorial and mathematical methods of representation of relief have been explained in detail with suitable examples and illustrations. Various types of distributional thematic maps have been analyzed with suitable examples emphasizing their suitable data structure, necessary numerical calculations, methods and principles of their construction, proper illustrations and advantages and disadvantages of their use. It also explains different techniques of measurement of direction, distance and area on maps.

The methods of construction of all types of graphs, diagrams and maps are explained step-by-step in a systematic way for easy and quick understanding of the readers. The book is unique of its kind as it reflects an accurate co-relation between the theoretical knowledge of various geographical events and phenomena and their realistic implications with suitable examples using proper graphical techniques. The book will be helpful for the students, researchers, cartographers and decision-makers in representing and analyzing various geographical data for a better, systematic and scientific understanding of the real world.

Midnapore, West Bengal, India   Swapan Kumar Maity

Acknowledgements

It gives me immense pleasure to express my deep gratitude to all those who contributed in their own ways for the successful completion of this book. I am heartily thankful to each soul that has come across all through the journey.

I owe my thankfulness to my students Rajesh Bag, Gopal Shee, Baneswar Adak, Suvajit Barman (SACT, K. D. College of Commerce and General Studies), Krishnapriyo Das and Arpita Routh for their help and support in preparing this book.

I would like to express my gratitude to Dr. Samit Maiti, Assistant Professor of English, Seva Bharati Mahavidyalaya and Dr. Soumitra Chakraborty, Assistant Professor of English, Mallabhum Institute of Technology for their academic support and advice. I am really thankful to Mr. Titas Aikat, GIS Manager, Horizen, Naihati and Mrs. Somrita Sinha, SACT, Department of Geography, Raja N. L. Khan Women's College for their technical and academic support for preparing this book. I am also thankful to Dr. Netai Chandra Das, Officer-in-Charge and Assistant Professor of Philosophy, Nayagram P. R. M. Government College for his continuous encouragement and valuable suggestions.

I would like to express my heartfelt gratitude to Dr. Ramkrishna Maiti, Professor, Department of Geography and Environment Management, Vidyasagar University for his encouragement, support and advice. His valuable suggestions at the time of preparing this book helped me in bringing it to the final shape.

I convey a lot of thanks to all my family members, especially to my wife Sonali for her encouragement, co-operation and continuous moral and emotional supports. I am very much thankful to my little son Souparno for his co-operation, which has given me sufficient time for the completion of this book.

Midnapore, West Bengal, India   Swapan Kumar Maity

About This Book

Representation of geographical data using graphs, diagrams and mapping techniques is a key for geographers and for researchers in other disciplines to explore the nature of data, the pattern of spatial and temporal variations and their relationships and formulation of principles to accurately understand and analyze features on or near the earth's surface. These modes of representation also enable the development of spatial understanding and the capacity for technical and logical decision-making. The book depicts all types of graphs, diagrams and maps, explained in detail with numerous examples. The emphasis is on their appropriate data structure, the relevance of selecting the correct technique, methods of their construction, advantages and disadvantages of their use and applications of these techniques in analyzing and realizing the spatial pattern of various geographical features and phenomena. This book is unique in that it reflects an accurate correlation between theoretical knowledge of geographical events and phenomena and their realistic implications, with relevant examples using appropriate graphical methods. The book serves as a valuable resource for students, researchers, cartographers and decision-makers to analyze and represent various geographical data for a better, systematic and scientific understanding of the real world.

Contents

About the Author

Dr. Swapan Kumar Maity is an assistant professor of geography at Nayagram P. R. M. Government College, Jhargram, West Bengal, India. He completed his doctoral degree at Vidyasagar University with his dissertation titled Mechanisms of sedimentation in the lower reach of the Rupnarayan River. Dr. Maity has 7 years of teaching experience at the undergraduate level in geography and 2 years at the postgraduate level in geography and environmental management. His teaching interests include geotectonic, geomorphology, climatology and practical geography, including remote sensing and GIS. His main research areas include fluvial geomorphology, river sedimentation and sediment mineralogy. He has published several research articles in renowned journals and two books from Springer in the field of the mechanism and environment of river sedimentation. Dr. Maity is a life member of the Indian Institute of Geomorphologists.

Abbreviations

ADCP	Acoustic Doppler Current Profiler
AE	Actual evapotranspiration
BB	Backward Bearing
CI	Cropping Intensity
DMS	Degrees, Minutes, and Seconds
DSMs	Defence Series Maps
DST	Department of Science and Technology
EGM 96	Earth Gravitational Model 1996
EI	Erosional integral
FAO	Food and Agricultural Organization
FB	Forward Bearing
GCA	Gross Cropped Area
GCS	Geographic Co-ordinate System
GPS	Global Positioning System
GRS-80	Geodetic Reference System 1980
GSI	Geological Survey of India
GTS	Great Trigonometrical Survey
HI	Hypsometric integral
IMF	International Monitory Fund
IQR	Inter-quartile Range
ISI	Indian Statistical Institute
MSL	Mean Sea Level
NATMO	National Atlas and Thematic Mapping Organization
NCA	Net Cropped Area
NMP	National Map Policy
NRSA	National Remote Sensing Agency
OSMs	Open Series Maps
PE	Potential evapotranspiration
PWD	Public Works Department
QB	Quadrantal Bearing
RB	Reduced Bearing

RL	Reduced Level
SLR	Satellite Laser Ranging
SOI	Survey of India
UNO	United Nations Organization
USDA	United States Department of Agriculture
UTM	Universal Transverse Mercator
VLBI	Very Long Baseline Interferometry
WCB	Whole Circle Bearing
WGS 84	World Geodetic System 1984

Symbols

f_i	Class frequency
x_i	Class mark
w_i	Class width
f_{di}	Frequency density
Rf_i	Relative frequency
N	Total frequency
F	Cumulative frequency
r	Radial distance
θ	Azimuthal angle
ϕ	Polar angle or zenithal angle
δ	Latitude
P	Precipitation
T	Temperature
R	Soil moisture recharge
U	Utilization of water
D	Deficiency of water
S	Surplus of water
Q	Water discharge
G	Gini co-efficient
Q_1	Lower quartile
Q_2	Middle quartile
Q_3	Upper quartile
P_r	Population of r ranking city
P_1	Population of 1st ranking city
h_i	Mid-value of the contour height
H	Maximum height of the basin
a_i	Area between successive contours
A	Total basin area
S_k	Skewness
β_1	Skewness co-efficient
μ_3	Third central moment

σ	Population standard deviation
μ_1	First moment
$f(x)$	Probability density function
μ	Population mean
β_2	Kurtosis co-efficient
l_i	Length of side of equilateral triangle or square or cube
r_i	Radius of the circle or sphere
f	Flattening
e	Eccentricity of the ellipse
H	Topographic height or Orthometric height
h	Spheroid or ellipsoid height
N	Geoid height
I	Magnetic inclination or magnetic dip
Z	Vertical component
D	Map distance
S	Map scale
T	Total number of full squares

List of Figures

List of Tables

Chapter 1
Concept, Types, Collection, Classification and Representation of Geographical Data

Abstract Geography is a scientific discipline which emphasizes on the collection, processing, suitable representation and logical and scientific interpretation of various types of primary and secondary data for better understanding and explanation of the spatial distributions and variations of different geographical features and phenomena on or near the surface of the earth. This chapter focuses on the concept and types of data used in geographical analysis, sources of each type of data, methods of their collection as well as the advantages and disadvantages of their use. Major differences between various types of data are discussed clearly with suitable examples. It includes the detailed discussion of the concept of attribute and variable, types of variables and differences between them. Different types of measurement scales used in geographical analysis, their characteristics and application in geographical study have been explained with numerous examples. Techniques of classification, tabulation and processing of the collected data on different basis (i.e. based on location, time etc.) are discussed properly with special emphasis on the preparation of frequency distribution table and related terminologies. Methods of representation of all types of geographical data, their appropriateness and advantages and disadvantages have been explained with suitable examples.

Keywords Geographical data · Primary data · Secondary data · Data collection · Measurement scale · Data processing · Data representation

1.1 Introduction

Geography is a scientific discipline in which various types of *primary and secondary data* are used voluminously to explain and analyse different geographical events and phenomena. Collected data are organized, represented and interpreted logically and scientifically using various techniques for better understanding and explanation of the spatial distributions and variations of different geographical features on or near the surface of the earth. *Geographical data* needs appropriate, systematic and logical presentation for better understanding of their cartographic characteristics. Suitable, accurate and lucid demonstration and visualization of geographical data becomes helpful for their correct analysis, explanation and realization.

S. K. Maity, *Essential Graphical Techniques in Geography*, Advances in Geographical and Environmental Sciences, https://doi.org/10.1007/978-981-16-6585-1_1

Geographical studies emphasize how and why different features vary from one place to another and how spatial patterns of these features change with time. Geographers always begin with the question 'Where?', investigating how different features are located on a physical or cultural area, monitoring the spatial patterns and the variations of features. Modern geographical study has shifted to 'Why?', specifying why a particular spatial pattern exists, what kinds of processes (spatial or ecological) have influenced the pattern, as well as why such processes operate. Graphical visualization of various geographical data is the key to realize the nature and character of data, the pattern of their spatial and temporal variations, communicating the knowledge of spatial information, to classify different features and objects and understanding their relationships, formulation of principles which become helpful for proper understanding of the real world.

Graphs, diagrams and maps are three important methods of visual representation of data in geography. These three methods are unique and distinctive in terms of the principles and procedures followed and applied for their depiction. In narrow sense, graphical representation means portraying of data using various types of graphs but in broader geographical perspective, all types of graphs, diagrams and mapping techniques are considered as graphical methods of presentation of data. Different *graphical techniques (graphs, diagrams and maps)* are very popular to the geographers and researchers as they help for better understanding of the world around us by enriching spatial intelligence and capacity of human beings for technical and logical decision-making. Graphical representation of geographical data is very simple, attractive and easily recognizable not only to the geographers or efficient academicians but also to the common literate people.

1.2   Concept of Data

A body of information in numerical form is known as *data*. In other words, data are characteristics or information which are generally numerical in nature and are collected through observation. In technical sense, data means a set of values in quantitative or qualitative form concerning one or more individuals or objects. Data contain some facts and information from which an inference may be made or a reliable conclusion may be drawn. Actually, data are the raw materials of any type of research or investigation. So, the collection of reliable and dependable data is the prerequisite condition for conducting any research or investigation and drawing consistent conclusions.

For instance, if we want to analyse the trends and patterns of rainfall distribution and its changes over time of any region, at first daily or monthly rainfall data (in numeric figure) should be collected and then suitable techniques should be applied on those collected data for drawing reliable inferences.

Major characteristics of data are:

1.   Data must be represented in numerical forms.

2. All the data must be interrelated with each other.
3. They must be meaningful to the purpose for which they are required.

1.3   Concept of Geographical Data

Data that record the locations and characteristics of natural or human features or activities which occur on or near the earth's surface are called *geographical data*. Two important characteristics of these data are—i) the reference to geographic space or earth surface (expressed by geographical co-ordinates) and ii) the representation at specific geographic scale.

Example Amount of rainfall, temperature, height from mean sea level, number of landslide occurrences, volume of water discharge in river, number of population, density of population, production of agricultural crops etc. are considered as geographical data as they possess the above mentioned two characteristics.

Geographers and researchers use huge amount of statistical information and data for proper and logical understanding and explanation of various geographical features and phenomena on or near the earth's surface. Different statistical techniques and principles are widely used by them for the correct and scientific processing, analysis, depiction and interpretation of the collected data.

1.4   Types of Data (Geographical Data)

Like other data, geographical data are also of two main types on the basis of their nature and characteristics:

1.4.1   Qualitative Data (Attribute)

The qualitative characteristic of the information which can't be measured and expressed in numerical or quantitative terms is called *qualitative data or attribute*. Attribute refers to the characteristics of the quality of an observation which can be observed, ascertained and classified under different categories but can't be expressed in quantitative or numerical forms. There are numerous qualitative data which are used in geographical study.

For example, skin colour of the people, educational status, efficiency, caste system, attitude and mentality of people etc. are this type of data. All the qualitative data are converted into numerical or quantitative data for efficient and successful application of statistical techniques during geographical investigation. For instance, 100 people are literate, 150 people are of general caste, skin colour of 340 people is black etc.

In all those cases, the quality or characteristics of the data has been converted into numeric forms.

1.4.2 Quantitative Data (Variable)

The characteristic of the information which can be measured and expressed numerically or quantitatively in suitable units is called *quantitative data or variable*. Variable refers to the quantitative characteristic of an individual or item which takes different values depending on situation and place and these values can always be measured. The variable whose values depends on chance and can't be predicted is called random variable.

For example, average monthly rainfall is 15 cm, number of first order streams in the river basin is 345, average volume of water discharge in the stream is 560 m³/sec, rate of soil erosion is 1 mm/year, production of rice is 1200 kg/acre, literacy rate of the country is 65%, fertility rate of any country is 12 persons/year/1000 peoples etc. are the quantitative expression of data. These data are more suitable for the application of statistical techniques and successfully used in geographical analysis.

1.4.2.1 Continuous Variable and Discontinuous or Discrete Variable

The variable which can take any value within a specified range is called *continuous variable*. These variables can be expressed not only in integral part, but also in fraction of any part, however small it may be. When the continuous variables are represented in the form of a series, then it is known as continuous series.

Example Amount of rainfall, temperature, height from sea level, velocity of river water, literacy rate, weight of people etc. are the examples of continuous variables. The amount of rainfall may be 25 cm or 25.5 cm or 25.55 cm or any other values. Similarly, the literacy rate may be 68% or 68.25% or 68.59% or any other values.

The variable which can assume only some isolated values or integral values is called *discontinuous or discrete variable*. Discrete variables can't be expressed in fractional values. When the discontinuous or discrete variables are expressed in the form of a series, then it is called discontinuous or discrete series.

Example The number of streams in different orders in a river basin, number of household in a village, number of agricultural or industrial workers in a country, number of peoples affected by flood hazard, number of migrated peoples etc. are the examples of discrete variables. The number of household in a village may be 205 or 206. But, it can't be 205.65 as a household can't be divided into parts or fractions. Similarly, the number of people affected by flood hazard may be 450 or 451, but it can't be 450.5 or 450.75.

1.4.2.2 Difference Between Continuous and Discontinuous or Discrete Variables

Major differences between continuous and discontinuous or discrete variables are as follows:

Continuous variable	Discontinuous or discrete variable
1. These variables can take any value within a specified interval	1. These variables can take only some isolated or integral values
2. Can be expressed not only in integral part, but also in fraction	2. Can be expressed only in whole numbers, fractional expression is not possible
3. These variables are measurable but not countable	3. These variables are countable but not measurable
4. Continuity of representation of variables is maintained	4. Continuity of representation of variables is not maintained
5. Variables are expressed by a range, like $\infty \leq X \leq \infty$	5. Variables are expressed by a fixed value, like $X = 0, 8, 12, 15$, etc
6. **Example:** Height, rainfall, temperature, velocity etc.	6. **Example:** Number of households, number of students, number of accidents etc.

1.4.3 Uni-Variate Data and Bi-Variate Data

Statistical data relating to the measurement of one variable only are called *uni-variate data*. For example, amount of organic matter in soil, concentration of Suspended Particulate Matter (SPM) in air, annual production of rice, income of a family etc. Generally, central tendency, dispersion, skewness and kurtosis etc. are used as the statistical measurements of these variables. Uni-variate data are represented by a letter or symbol 'x' and the 'n' number of values of 'x' variable are expressed by $x_1, x_2, x_3, x_4, \ldots \ldots \ldots \ldots x_n$.

Data relating to the simultaneous measurement of two variables are called *bi-variate data* (Table 1.1). For example, height from mean sea level and number of settlements, volume of surface run-off and rate of soil erosion, income and expenditure of a family, amount of fertilizer used and crop production, distance from Central Business District (CBD) and lower atmospheric temperature etc. Here, one variable is influenced by another variable and thus bi-variate data has an independent and a dependent variable. Co-relation and regression are popular statistical techniques for the analysis of these variables. Bi-variate data are represented by two letters or symbols (x_i, y_i) and the 'n' pairs of values are expressed by (x_1, y_1), (x_2, y_2), (x_3, y_3) .... (x_n, y_n).

Table 1.1 Bi-variate data showing depth below ground (m) and air temperature (°C)

Sl. No	Depth below ground (m)	Air temperature (°C)	Sl. No	Depth below ground (m)	Air temperature (°C)
1	0	10.6	9	840	22.1
2	140	11.6	10	690	22.6
3	300	13.3	11	590	23.6
4	170	13.8	12	820	25.5
5	310	15.1	13	1020	26.9
6	340	17.0	14	1150	30.2
7	460	19.3	15	970	30.6
8	550	20.6	16	830	26.2

1.4.4   Difference Between Uni-Variate Data and Bi-Variate Data

Major differences between uni-variate and bi-variate data are as follows-

Uni-variate data	Bi-variate data
1. The word 'Uni' means one. Statistical data involving one or single variable is called uni-variate data	1. The word 'Bi' means two. Statistical data involving two variables (one independent and one dependent variable) is called bi-variate data
2. It is not associated with causes or relationships	2. It is closely associated with causes or relationships
3. Description of a specific variable is the main purpose of uni-variate analysis	3. Explanation is the main purpose of bi-variate analysis
4. Central tendency (mean, median and mode), dispersion (range, quartile, mean deviation, variance and standard deviation), skewness and kurtosis are the main techniques of uni-variate analysis	4. It uses different techniques like co-relations, regression, comparisons, causes and explanations etc. for the analysis of two variables simultaneously
5. The result of uni-variate analysis is shown in bar graph, pie-chart, line graph, box and whisker plot etc.	5. The result of bi-variate analysis is shown in table where one variable is contingent on the values of the other variable
6. **Example:** Annual production of rice, annual precipitation, amount of suspended particulate matter in air etc.	6. **Example:** Relation between height from mean sea level and number of settlements, volume of surface run-off and rate of soil erosion, distance from Central Business District and temperature etc.

1.4.5 Independent Variable and Dependent Variable

The variable which stands alone and does not depend on other variables, moreover controls other variables is called *independent variable*. Independent variables may be one or more in number. This variable is expressed by 'x' and is shown along 'X'-axis in graph.

The variable which depends on other variables and is affected by them is called *dependent variable*. The value of dependent variable undergoes changes due to the change of value of independent variables. This variable is expressed by 'y' and is shown along 'Y'-axis in graph.

Example In the above data (Table 1.1), air temperature changes with the change of depth below ground. So, the air temperature is dependent variable and the depth below ground is independent variable. Again, the production of agricultural crops depends on availability of water, amount of fertilizer used, labour used etc. Here, crop production is dependent variable but availability of water, amount of fertilizer used, labour used are independent variables.

1.4.6 Difference Between Qualitative Data (Attribute) and Quantitative Data (Variable)

The following are the differences between qualitative and quantitative data:

Qualitative data (Attribute)	Quantitative data (Variable)
1. Data representing the qualitative characteristics of the statistical information	1. Data representing the quantitative characteristics of the statistical information
2. Data can be observed, ascertained and classified under different categories but can't be expressed in numerical form	2. Data attain different values which can easily be measured and expressed in numerical form
3. Data should be transformed into quantitative forms before used in statistical analysis	3. Different statistical techniques can easily be applied on those data
4. **Example:** Educational status, caste system, attitude etc.	4. **Example:** Amount of rainfall, volume of water discharge, rate of sediment transport etc.

On the basis of sources of collection, geographical data are of two types.

1.4.7 Primary Data

Primary data are those data which are collected for a specific purpose directly from the field of investigation, and hence are original in nature. These types of data are collected originally by the individual, group or authority who requires the data for their own use and treatment. These data have not been used in quantitative research previously. These are called raw data or basic data as they are directly collected from the field by the field-workers, investigators and enumerators. The level of accuracy and reliability of these data depend on the knowledge, efficiency, consciousness and mentality of the researcher or investigator and also on the methods of data collection. The places or sources from which primary data are collected are known as primary sources.

Examples The data collected from measurements of river depth, width, water velocity, water discharge, tidal water level etc. directly in the field by the researcher using various instruments are primary data. Similarly, various socio-economic data (caste, religion, literacy rate, job opportunity, income, expenditure, marital status, immunization status etc.) collected from household survey using survey schedule by the researcher are the examples of primary data.

1.4.8   Secondary Data

The data which have previously been collected and published by someone for one purpose but subsequently treated and utilized by another one in a different connection are called *secondary data*. Secondary data are actually collected and published by the organizations other than the authorities who need them subsequently for their use. So, primary data of one organization become the secondary data of other organization who later want to use those data. Because of this, secondary data are not considered as basic data. The sources from which secondary data are collected are known as secondary sources.

Examples Data, collected from any published books and journals, from different maps, from internet sources, from different government and non-government offices are the examples of secondary data. The Statistical Abstract of India and Monthly Abstract of Statistics, published by Central Statistical Organization and other publications of Government are different sources of secondary data.

1.4.9   Advantages of Use of Primary Data Over the Secondary Data

There is no hard and fast rule about which data should be used in *geographical research or investigation*. The nature, scope and purpose of the geographical enquiry should be taken into consideration whether primary data should be used or secondary data are to be utilized. Though the utilization of secondary data is more convenient and economical, but the use of primary data is preferable and much safer from several standpoints:

(a) Primary data usually show more detailed information and a description of the investigation along with the unit of measurement.

(b) The methods, sources and any approximations used for the collection of data are clearly and specifically mentioned in those data. So, it can be decided in advance how much reliance can be given on those data while they are being used.

(c) Primary data are more reliable, authentic and accurate than secondary data as the later contain errors because of transcription, rounding etc.

Inspite of this, the secondary data are used due to the following reasons:

(a) Primary data are not available or can't be collected directly due to limitations of time and money during data collection.

(b) To compare the data collected over a long period of time, the use of secondary data is required. Utmost accuracy is not so much necessary in these cases.

1.4.10 Difference Between Primary and Secondary Data

Actually, *primary data* and *secondary data* are same because the former is transformed into the later with the advancement of time. The major differences between primary and secondary data are:

Primary data	Secondary data
1. Primary data are collected directly from the field or area under study by the investigator	1. Secondary data are collected from any published books or journals, offices, internet sources, institutions etc.
2. Data are the result of direct observations and interactions in the study area	2. Data are mainly the result of publications
3. Trained and efficient manpower is needed during the collection of primary data	3. Trained and efficient manpower is not essential for the collection of data. Non-trained person can collect the data
4. Quality of data is largely affected by the knowledge, efficiency, consciousness and mentality of the investigator	4. Researcher or investigator has no role to control the quality of data
5. Data are more accurate, authentic and reliable	5. Data are less reliable due to the possibility to be erroneous
6. Data are always collected in original unit	6. Data can be collected in original unit or in any other converted unit, like aggregate, ratio, average, percentage etc.
7. Collection of data is time consuming, costly and sometimes becomes risky	7. Data collection is less time consuming and cost effective
8. These data are at the first stage of their utilization and numerical techniques are not applied previously	8. Different numerical techniques have been applied previously in those data, i.e. they are in second, third or any other stages of their utilization
9. Primary data is preferred more by the researchers in statistical investigations because of its several advantages	9. Secondary data is used in those cases when primary data is unavailable or can't be collected directly

1.5   Methods of Data Collection

There is no hard and fast rule in adopting a specific method for the collection of geographical data. The method of data collection is decided by the objectives and purposes of the study. Young (1994) has divided the data sources into two classes: (a) field sources and (b) documentary sources. Field sources are the sources of primary data whereas the documentary sources are the sources of secondary data.

1.5.1   Methods of Primary Data Collection

Generally, five methods are followed for the collection of primary data:

1.   Observation method
2.   Interview method
3.   Sampling method
4.   Experimentation method
5.   Local sources method.

1.5.1.1   Observation Method

Continuous and intensive observation of different objects, events or phenomena is an important method of primary data collection. The success of this method depends on the knowledge, efficiency and capability of the researcher or investigator. There are three types of observations:

Direct Observation Method

The researcher or investigator collects the necessary information directly by himself or herself being present in the field. The researcher visits the area to be studied keeping some hypotheses in his/her mind. After intensive and careful field observation, some new ideas and experiences are added to the previous hypotheses which help to develop the theory and make the collected *primary data* reliable and relevant.

Example In case of landslide study, the researcher or investigator directly visits the land slide affected area and collects different data regarding the total area affected by land slide, volume of materials displaced, length of sliding, slope of the land, composition of materials etc.

Advantages and Disadvantages of Direct Observation Method
Advantages

(i)   More reliable and accurate data can be collected without any biasness.
(ii)   Usable for small area investigation.

(iii)   Privacy of data can be maintained.
(iv)   Clarity and homogeneity of data can be maintained.
(v)   Collection of complete data is possible.

Disadvantages

(i)   Probability of wastage of time and money.
(ii)   Method can't be applied in large study area.
(iii) Sometimes, the self-feelings, emotions, mentality and prejudices of the researcher affect the collection method and quality of the data.
(iv)   Sometimes, the data collection becomes risky.

Indirect Observation Method

When the responder is not agreeing to provide information or to answer the questions accurately, then this method is applied. In this situation, the responder is avoided and information is collected from the associated third person. Data is collected by the researcher himself or herself or by the enumerator appointed by the researcher.

Advantages and Disadvantages of Indirect Observation Method
Advantages

(i)   Less time consuming and cost effective.
(ii)   Effective for the collection of qualitative data.
(iii) Data can be collected in risky condition.
(iv)   Effective for data collection in large population.
(v)   Unbiased data collection is possible.

Disadvantages

(i)   The data are not as reliable as collected from the associated third person.
(ii)   Information provider may be biased.
(iii) Data may be biased due to negligence of information provider.
(iv)   Collected data may be erroneous due to lack of trained enumerator.

Participation Observation

The researcher or investigator collects the information by staying, living and inter-acting with the people of the area under study. In this method, the researcher makes a close and intricate relationship with the local people of the area and observes their daily activities and life style intensively. Questions are not asked to the people but data are collected by observations, feelings and individual judgements of the investigator.

Example For the intensive study of the livelihood pattern and social adjustments of the people of any indigenous tribal society, the researcher or investigator live and makes a close relationship with the people of the society for collecting required information for the fulfilment of the purpose.

Advantages and disadvantages of participation method
Advantages

(i)   Reliable, unbiased and accurate data are collected as the researcher collects the data by making close relation with the local people.
(ii)   Simple, easy and unambiguous technique of data collection.
(iii) Effective in qualitative data collection.
(iv)   Collection of data about any specific group of people becomes possible.
(v)   Researcher can change and modify the hypothesis of research easily.

Disadvantages

(i)   Complete observation and understanding about the research area is difficult as it is totally unknown to the researcher.
(ii)   It is a valiant and risky method for data collection.
(iii) Time consuming and costly because the researchers have to stay in the research area for a certain period of time.
(iv)   Prior experience and training is required for the collection of data.
(v)   Limited applicability in large research area.

1.5.1.2   Interview Method

In *interview method*, information is collected by the conversation between investigator or enumerator (interviewer) and the informant (interviewee). The interviewer makes a close interaction and face-to-face discussion with the informant for collecting the data. Interview methods are of three types.

Interviewing by Questionnaire Method

In this method, the enumerators interview the concerned persons directly or indirectly and ask questions to collect information. The information is gathered generally on standard set of questions. Before collecting the data, a standard *questionnaire* is prepared by the researcher.

Example If a researcher wants to study the impact and management of flood in any flood-prone area, he/she will prepare a standard questionnaire considering the following points, like causes of flood, frequency of flood, duration of water logging during flood, area affected by flood, problems of flood, sources of food and drinking water during flood, flood controlling measures taken by governmental and non-governmental agencies, precautions to avoid flood, any advantages from flood etc.

Characteristics of Standard Questionnaire

(a)   The questions should be meaningful, concise, clear and easy to understand to the interviewee.

(b)   The number of questions should be limited and they will be arranged sequentially and systematically.

(c)   Questions should be impartial and unbiased to avoid the hesitation of the interviewee.

(d)   All the questions should be relevant and will be sufficient for the fulfilment of the purpose of the research.

(e)   Questions should be free of religious, political and other prejudices.

(f)   Calculative questions should be avoided.

Questionnaire method is of two types.

Direct Questionnaire Method

In this method, the researchers or the enumerators appointed by the researcher go personally to the persons or to the sources from whom (which) the information should be collected. The enumerators interview the concerned persons and ask the questions directly (face to face) during the time of data collection. This method is also called interview schedule method.

Advantages and disadvantages of Direct Questionnaire method
Advantages

(i)   Data confirm high degree of accuracy as the investigators or enumerators have direct contact with the people (interviewee).

(ii)   Data are more reliable and dependable.

(iii) The purpose of the study and the meaning of each question can clearly and patiently be explained to the interviewee.

(iv)   It helps to collect the relevant information only.

(v)   Privacy of data will be maintained.

(vi)   Testing of data accuracy is possible.

Disadvantages

(i)   It is very expensive, time consuming and complex technique of data collection.

(ii)   Difficult to apply for large observations in extensive area.

(iii) Untrained and inefficient enumerator may collect erroneous information.

(iv)   It allows the personal prejudices of the enumerators or the investigators to affect the quality of the data and the inferences to be drawn.

Postal Method of Questionnaire Survey

A standard questionnaire is prepared and sent to different addresses by post for answering the questions. Generally, all the questionnaires are accompanied by a letter of explanation and self-addressed envelopes in order to receive the information properly at the earliest. This method is widely used for data collection in planning process.

Advantages and disadvantages of Postal Questionnaire method
Advantages

(i)   This method helps extensive investigations and covers the large fields of study.
(ii)   It is a very easy and quick method. Data can be collected within very short time.
(iii)   It is cost effective for data collection. Only postal charges are required.
(iv)   It is free from personal bias of the enumerators or investigators.
(v)   Data can be easily collected from long distances.
(vi)   Very useful to judge the national point of view.

Disadvantages

(i)   Some of the questionnaires may not be answered and returned to the researcher.
(ii)   Questionnaires may be returned to the researcher without giving proper answer and filled in. Wrong answer may be given without understanding proper meaning of the questions.
(iii)   Method can't be applied to the informants who are illiterate or who are ignorant about the importance and requirement of the information.
(iv)   The accuracy of the information can't be verified. So, data are not so much reliable and dependable.

Interviewing by Informal Method

In this method, the researcher or investigator collects the required information out of the inadvertence of the informants. Generally, this method is applied for the collection of information about any specific phenomenon or event. When the informants are not agreeing or hesitating to provide sufficient information, then the investigator attempt to collect necessary information by immaterial and superfluous discussion with the informants. The informants explain the actual fact unintentionally to the investigator which becomes important information to the researcher. No questionnaire is needed for collecting data by this method. The researcher or investigator collects the data by asking the questions to the informants from his/her memory.

Example For conducting a study about the dimension and status of illegal coal mining in any region, the researcher needs to collect information regarding the area and number of illegal coal mines, number of people engaged in this work, amount of daily coal withdrawal, means of mining, major uses of the mined coal, any problems faced by the miners etc.

Advantages and disadvantages of Informal interview method
Advantages

(i)   Easy data collection by this method by the extrovert persons.
(ii)   Qualitative data can easily be collected by this method.
(iii)   Collected data are more reliable and dependable.
(iv)   Behaviour of the informants is expressed accurately.

Disadvantages

(i) Trained and efficient investigator is required.
(ii) Consciousness of the enumerator or investigator is essential.
(iii) Very costly and time consuming.

Interviewing by Telephone

The researcher collects the information by telephonic interview of the informants. When enormous information is needed urgently within very short time, then this method is followed.

Advantages and disadvantages of Telephone interview method
Advantages

(i) Collection of huge information within very short time and spending little money.
(ii) Lesser number of investigators is needed.
(iii) Data are reliable and collected systematically.
(iv) Very suitable to apply in small research area.

Disadvantages

(i) Applicability of this method depends on the availability of telephonic communication.
(ii) Informants are not always available in urgent condition.

1.5.1.3 Sampling Method

Sampling is a very important method for the collection of different primary data. Reliable statistical inferences can easily be drawn about a large number of observations (population) under study by testing small samples collected from the population. The members of the population which are selected for statistical testing are called samples and the technique of sample selection is called sampling. Sampling technique is used very popularly and significantly for the collection of data to be used in different geographical study and research.

Example For assessing the quality of water of a lake, required number of water samples should be collected from different parts and depths of the lake.

Similarly, the socio-economic status of the people of a large slum area should be studied by collecting required data from the slum households. When the collection of data from all slum households is not possible due to time and money constraint, then slum households should be selected by suitable sampling method.

Advantages and Disadvantages of Sampling Method
Advantages

(i)   Data collection is cost effective and less time consuming.
(ii)   It is applicable for all types of geographical survey.
(iii) Minimum number of investigators is required.
(iv)   All the factors regarding survey and data collection can be monitored carefully.
(v)   Trained and efficient researcher can solve different critical problems by collecting data in this method.

Disadvantages

(i)   Collection of data by sampling technique requires trained and efficient enumerator or investigator.
(ii)   Presumptions, prejudices, partiality and negligence of the enumerator or investigator will affect the selection of samples.
(iii) Wrong sampling technique or collection of wrong samples makes the result erroneous and less applicable.
(iv)   Results from sample study may not always reflect all the characteristics of the whole population.

1.5.1.4   Experimentation Method

It is an important part of the sampling method for primary data collection. In this method, the researcher or investigator collects the required samples from the study area, analyse and test the collected samples in the laboratory or research centre and generates numerous primary data.

Example For knowing the mineral composition of soils of any region, we have to collect the required number of soil samples from the concerned study area, and then the collected samples should be tested and experimented in the laboratory using different instruments (preferably using X-Ray Diffraction technique) and chemicals.

Similarly, if we want to know about the arsenic contamination of groundwater of any region, then sufficient numbers of groundwater samples should be collected from wells, tube-wells or any other sources for testing them in the laboratory to generate primary data.

Advantages and Disadvantages of Experimental Method
Advantages

(i)   It is an ideal method for generating data in the laboratory.
(ii)   Numerous data can be generated within very short time.
(iii) All types of variables can be controlled and monitored easily.
(iv)   It needs minimum number of enumerator or investigator.

Disadvantages

(i)   Generation of data in this method is very costly.

(ii)   It can't be applied in all types of geographical research.
(iii)   Trained and efficient investigators are required for the utilization of different instruments and peripherals in the laboratory.

1.5.1.5 Local Sources Method

The researcher or any institution appoints the local people of the research area as the enumerator or investigator for collecting data about any geographical phenomenon or event. Being local residents, the enumerator comprise clear-cut and explicit idea about the study area. They collect all the required information by direct observations about any phenomena and send the collected data to the concerned researcher or institution. Generally, different types of regional geographical data are collected by this method.

Example Measurement of daily water discharge of a stream, measurement of hourly tidal water level in a tidal river, collection of weather-related data (atmospheric temperature, amount of rainfall, wind direction, wind velocity, air pressure, humidity etc.) at any weather station should be done by appointing the local residents as the investigator.

In socio-economic survey, the study of daily livelihood pattern of a particular group of people can be made following this method.

Advantages and Disadvantages of Local Sources Method
Advantages

(i)   Data can be collected continuously and instantly.
(ii)   Fewer enumerators can collect the required data.
(iii)   Decision can be made quickly.
(iv)   Place-wise data can be collected from an extensively large research area.

Disadvantages

(i)   Method of data collection is very costly.
(ii)   Untrained and inefficient investigator can collect erroneous information.
(iii)   Partiality and prejudices of the investigator deteriorates the quality of data.
(iv)   Sometimes, the data are collected based on assumptions which makes the data undependable.

1.5.2 Methods of Secondary Data Collection

There are no proper methods for the collection of secondary data. Generally, secondary data are collected from two main sources:

1. Published sources
2. Unpublished sources.

1.5.2.1   Published Sources

Numerous secondary data are collected from the published reports, records and documents of government offices and other non-government departments and agencies. Government and non-government departments and agencies prepare and publish different reports, records and documents on various subjects. Data are often collected by the researcher or investigator from those published sources.

Example The main sources of collecting data under this method are (a) publications of government, (b) reports of different commissions and committees, (c) reports and publications of trade associations and chambers of commerce, (d) market reports and business bulletins of stock exchanges, (e) economic, commercial and technical journals, (f) publications of researchers and research institutions etc. Secondary data are published by different organizations like United Nations Organization (UNO), International Monitory Fund (IMF), Food and Agricultural Organization (FAO), World Bank, UNESCO, UNICEF, Indian Statistical Institution (ISI) etc. In India, National Remote Sensing Agency (NRSA), SIO, National Atlas and Thematic Mapping Organization (NATMO), Geological Survey of India (GSI), IMO, SSI etc. are important sources of various geographical maps and data.

1.5.2.2   Unpublished Sources

Sometimes, the collected primary data are not properly published by the collector, called unpublished data. The researcher collects those unpublished data for their own need from the collector individual or institution through personal connection and relationship. For example, unpublished thesis paper of a scholar, records and documents stored in different governmental and non-governmental offices etc. are unpublished sources of secondary data.

1.5.2.3   Advantages and Disadvantages of Secondary Data Collection

Advantages

(i)   It helps in furnishing reliable information and reliable data.
(ii)   This method of data collection is inexpensive. The cost of data collection is borne by governmental and other non-governmental departments, offices and agencies.

Disadvantages

(i)   The unit of the published data may not be same as it was in the collected primary data. So, the data collected from published sources may not serve the purposes.

(ii) The basis of classification and the method of collection of data may also be different in governmental and non-governmental sources from which the secondary data are collected. Due to this, the data may not be appropriate for the fulfilment of the purpose of the researcher.

Detailed and careful scrutiny and verification of the data before putting them into use is the prerequisite condition of this method. The researcher or user of the data should scrutinize the data cautiously in order to know whether the data are appropriate for the purpose for which they are intended. Before the collection of those data, the following points may be taken into consideration: (1) the scope and objectives of the study for which the data were actually procured (2) units of the collected primary data (3) methods adopted for the collection of data (4) degree of authenticity and accuracy of the data (5) honesty and reliability of the authorities who collected the data.

1.6 Measurement Scales in Geographical System

Measurement can be defined as assigning the names and quantifying the earth surface features and working out the relationship using them. In general, measurement refers to the quantitative description (numerical value) of some properties or attributes of objects or events for comparing one object or event with others. It offers a platform to describe the attributes and to communicate this description with others. Measurements or data are the raw materials of descriptive and inferential statistics with which statistical techniques do work. Data includes facts or figures recorded as an outcome of measuring or counting a system and from which reliable inferences are made.

After the procurement and recording of the data regarding spatial or temporal distribution of any phenomenon or event or object, it needs to be properly categorized and summarized in numerical forms. This method of categorizing the collected raw data involves four different processes of measurement providing four types of 'number scales'. These are:

(a) Nominal scale
(b) Ordinal scale
(c) Interval scale
(d) Ratio scale

1.6.1 Nominal Scale

It is the basic and simple form of measurement in which data are expressed in terms of identity only like male or female, lowland or highland, unreserved or reserved category, present or absent etc. So, the *nominal scale* is similar to the binary scale in which the presence of any character or phenomenon is expressed by the value '1' and the absence by '0' (Pal 1998). Tossing of a coin which gives either head or tail is the classical example of a nominal scale.

1.6.1.1 Characteristics of Nominal Data

a. Data should be exhaustive (includes all events or phenomena under study) and mutually exclusive (no value is laid in two or more category).
b. The items in each category are counted and the total is represented by a number.
c. Data can't be manipulated by any basic mathematical operation (addition, subtraction, multiplication, division etc.).
d. It is termed as count data in the form of frequencies.
e. All observations or items within each category are treated as same.

1.6.1.2 Application in Geographical Study

It is used for the determination of equality or differences between geographical phenomena or events. 'Mode' is used only as the measurement of central tendency in nominal data. Frequency, binomial and multinomial expression is easy in this type of data.

Examples Classification of land use pattern (forest land, cultivated land, built-up land etc.); soil, rock or mineral classification etc. belong to nominal scale. Table 1.2 shows the number of male and female students in different departments as an example of nominal scale measurement.

1.6.2 Ordinal Scale

It is the level of measurement superior to nominal scale. In this method, there is sufficient information to place the data before or after another along a scale in rank order either individually or in groups. The differences between objects or events by their identities can easily be established by this method. The statement $X < Y < Z$

Departments	Number of students	
	Male	Female
Mathematics	35	15
Statistics	28	22
Physics	32	18
Chemistry	29	21
Geography	27	23

Table 1.2 Nominal data (Number of male and female students in different departments)

indicates that there are three values or classes of any object or phenomenon in which value or class X is less than value or class Y and again Y is less than value or class Z.

1.6.2.1 Characteristics of Ordinal Data

a. The direction and relative position of values on this scale are known.
b. The differences between objects or events by their identities can easily be established.
c. Application of mathematical operation (addition, subtraction, multiplication, division etc.) is not possible.
d. The actual differences between values can't be understood.
e. Some data are inherently ordinal in nature.

1.6.2.2 Application in Geographical Study

It is applied for the determination of greater or lesser values of observations related to any geographical phenomena or events, i.e. rank of different values of any observation can easily be identified in this scale. Mode, median, percentile and inter-quartile range (quartile deviation) are widely used as the measurement of central tendency and dispersion of values.

Examples Classification of families of any region into rich, upper class, upper-middle class, middle class, lower-middle class and lower class according to their socio-economic status is an example of ordinal scale. Similarly, the ranking of Indian states according to the literacy rate of the people is done using this scale (Table 1.3). Moh's scale of hardness of minerals is another example of ordinal scale (Table 1.4).

1.6.3 Interval Scale

Interval scale consists of measures for which there are equal intervals between each measurement or each group. Thus, interval scales are numeric scales in which not

Table 1.3 Ordinal data (Literacy rate of few Indian states, 2011)

Name of states	Literacy rate (%)	Rank	Name of states	Literacy rate (%)	Rank
Kerala	93.91	1	Maharashtra	82.91	6
Mizoram	91.58	2	Sikkim	82.20	7
Tripura	87.75	3	Tamil Nadu	80.33	8
Goa	87.40	4	Nagaland	80.11	9
Himachal Pradesh	83.78	5	Manipur	79.85	10

Table 1.4 Moh's scale of
hardness of minerals (Ordinal
scale)

Hardness	Mineral	Hardness	Mineral
1	Talc	6	Feldspar
2	Gypsum	7	Quartzite
3	Calcite	8	Topaz
4	Fluorite	9	Corundum
5	Apatite	10	Diamond

only the objects are given identities and ranked like nominal and ordinal scales but also the exact differences or intervals between objects in terms of their property are known. This is capable of comparing the differences between a number of pairs or values to specify the exact location of the objects along a continuous scale.

1.6.3.1 Characteristics of Interval Data

a. Direction and magnitude of position on scale are known.
b. The exact difference between any two values on the scale is known but there is an arbitrary point and a unit of measurement.
c. The interval scaled data can easily be added or subtracted but multiplication or division is not possible.
d. It represents precise idea about all the values of the data.
e. In this scale, the value of zero is arbitrary; absolute zero (true zero) is not used. For example, the zero (0) value in p^H scale is arbitrary.

1.6.3.2 Application in Geographical Study

It is used for the determination of equality or differences of intervals of the values in arithmetical sense. Interval scales are very applicable as the area of statistical analysis of geographical data sets opens up. Central tendency can be measured by mean, median or mode; variance, mean deviation and standard deviation can also be used as the measure of dispersion.

Examples Celsius temperature scale is the standard example of an interval scale as the difference between each value is same. The difference between 50 and 40 degree Celsius is a measurable 10 degree Celsius, as is the difference between 90 and 80 degree Celsius. Time is another classical example of interval scale in which the increments are recognized, consistent and measurable. For instance, time in years AD or BC. Longitude, compass directions are also the examples of interval scale.

1.6.4 Ratio Scale

All the requirements of the interval scale are met in *ratio scale,* and in addition, it has an absolute zero scale (Pal 1998). For instance, rainfall scale (either in centimetres or in inches) has a true zero base. Thus, if a place 'A' receives 60 cm rainfall and place 'B' receives 180 cm rainfall in a year, we can conclude that the place 'B' receives three times more rainfall in a year than the place 'A'. Ratio scale tells us about the order of values, exact value between units and allows for a wide range of application of both descriptive and inferential statistics.

1.6.4.1 Characteristics of Ratio Data

a. Two measurements bear the same ratio to each other independent of the units of measurement.
b. Data are amenable to all types of mathematical operations (addition, subtraction, multiplication and division) and to many forms of statistical analysis.
c. Because of its absolute zero, the ratio scale contains maximum amount of information about any entity.
d. All the ratio variables are also interval variables but all interval variables are not necessarily ratio variables.

1.6.4.2 Application in Geographical Study

Ratio scale offers a wealth of possibilities when statistical data and techniques are used in geographical analysis. It is applied for the determination of equality or differences of ratios of the values. Central tendency can be measured by mean, median or mode; different measures of dispersion, for example, standard deviation and coefficient of variation can also be easily computed from ratio scales.

Examples Good examples of ratio scales are the measurement of height, weight, length, stream water velocity, slope, income of people etc. In all these measurements, zero point is identical and absolute.

The major characteristics of these four types of scale of measurement are shown in Table 1.5.

1.7 Processing of Data

Processing of data is very important and the prerequisite condition for the representation, analysis, explanation and interpretation of the collected data. The main aim of data processing is to make the data simple and comprehensible to all. Processing

Table 1.5 Characteristics of different scales of measurement

Characteristics	Nominal scale	Ordinal scale	Interval scale	Ratio scale
The order of values is known	No	Yes	Yes	Yes
Counts or frequency distribution	Yes	Yes	Yes	Yes
Mean	No	No	Yes	Yes
Median	No	Yes	Yes	Yes
Mode	Yes	Yes	Yes	Yes
Quantification of difference between each value	No	No	Yes	Yes
Addition or subtraction of values	No	No	Yes	Yes
Multiplication or division of values	No	No	No	Yes
Absolute or true zero	No	No	No	Yes

of data is nothing but the classification, arrangement and summarization of data. Galtung (1968) mentioned that 'Processing of data refers to concentrating, recasting and dealing with data such that they become as amenable to analysis as possible' (Khan 2006). The main procedure of data processing starts after the editing and coding of the data. Identification of errors in the collected data and their rectifications is called data editing. Soon after the collection of data starts, arrangements should be made to receive and verify the completed forms sequentially. Generally, many discrepancies, errors and omissions are observed in these completed forms. The defective forms should immediately be transferred back for necessary corrections. In case of plentiful errors and inconsistencies, the collected data should be cancelled and new data should be collected again. The completeness, uniformity, legibility and comprehensibility of the collected data should be checked carefully during the time of data editing. Data coding is executed after the editing of the data. Coding of data is the method of assigning numbers or symbols within the data. In close-ended question, data coding is performed before the collection of the data, but in open-ended question, data coding is done after the editing of the data.

Three methods are followed by the geographers and researchers for the processing of geographical data:

1.7.1 Classification of Data

The method of systematic arrangement of the data into different classes and groups based on their common characteristics and similarities is known as *data classification*. In the collected data, there is a group which have homogeneous and common characteristics and other groups of data are dissimilar from each other in terms of their characteristics. The homogeneous items are categorized into one group while the dissimilar items into another group. According to Kapur (1995), 'Classification is the process by which individuals and items are arranged into groups or classes according to their resemblances'. Good and useful classification of data should possess the unit of being exhaustive, mutually exclusive, stable and flexible. Data classification should also be specific and must not be ambiguous and clumsy.

1.7.1.1 Objectives of Data Classification

Major objectives of the classification of data are:

(i)   To ensure the sequential and systematic arrangement of data based on their characteristics, resemblances and affinity.
(ii) The nature, characteristics and actual conditions of the data should be understood and explained clearly by highlighting their similarity and dissimilarity in classification.
(iii) Simplification and summarization of data by reducing their complexities and ambiguities.
(iv) To make the data suitable for comparison and establishment of their relationship.
(v)   To make the data meaningful, comprehensible and easily applicable for depicting relevant inferences.

1.7.1.2 Characteristics of Ideal Data Classification

Though there is no hard and fast rule for the classification of data, but the following points should be taken into account during the classification of data:

(a) **Homogeneity:** Homogeneous and common values should be taken into one class and uncommon values into other class.
(b) **Purpose oriented:** Collected data should be classified in tune with the purpose of the research or investigation.
(c) **Clarity:** Classification should be clear, simple and easily understandable to all, complexities should be avoided.
(d) **Completeness:** All the items or values should be included in the classification carefully. No item should be eliminated during data classification.
(e) **Mutually exclusive:** Classes or groups should be mutually exclusive; no item should be included in more than one class.
(f) **Flexibility:** Though, stability of the data classification is important, yet the classification should be made in such a flexible way that further changes become possible.

1.7.1.3 Types of Classification

Generally, the classification of data is made based on the nature and characteristics of the collected data and the objectives of the study or investigation. There are four types of classification of data:

Table 1.6 Geographical classification of data (Population densities of some states in India, 2011)

Sl. No	Name of states	Population density (persons/sq. Km.)	Sl. No	Name of states	Population density (persons/sq. Km.)
1	Bihar	1102	6	Arunachal Pradesh	17
2	West Bengal	1029	7	Mizoram	52
3	Kerala	859	8	Sikkim	86
4	Uttar Pradesh	828	9	Nagaland	119
5	Tamil Nadu	555	10	Manipur	122

Table 1.7 Chronological classification of data (Decadal growth rate of population in India, 1901–2011)

Sl. No	Name of States	Decadal growth rate of population	Sl. No	Name of States	Decadal growth rate of population
1	1901	–	7	1961	21.64
2	1911	5.75	8	1971	24.80
3	1921	–0.31	9	1981	24.66
4	1931	11.00	10	1991	23.87
5	1941	14.22	11	2001	21.54
6	1951	13.31	12	2011	17.64

Geographical Classification (Based on Location or Space)

In this type, data regarding phenomena, events or objects are always measured and classified based on their geographical distribution and location. It is also called locational or spatial classification. For example, classification of state-wise production of rice in India, state-wise population density in India (Table 1.6), district-wise Scheduled Caste population in West Bengal etc.

Chronological Classification (Based on Time or Period)

In this type of classification, data are measured and arranged in sequence of time (chronologically) and classified according to the time by which the data are measured. The change of phenomena or events with respect to time is represented in this classification. For example, classification of month-wise water discharge in a river, year-wise total rainfall in India, decadal growth rate of population in India (Table 1.7), year-wise production of coal in India etc.

Fig. 1.1 Qualitative classification of data (population)

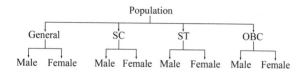

Table 1.8 Quantitative classification of data (Monthly income of a group of people)

Sl. No	Monthly income	Number of people	Sl. No	Monthly income	Number of people
1	Rs.5000–10000	25	6	Rs.30000–35000	14
2	Rs.10000–15000	28	7	Rs.35000–40000	21
3	Rs.15000–20000	18	8	Rs.40000–45000	12
4	Rs.20000–25000	30	9	Rs.45000–50000	10
5	Rs.25000–30000	45	10	Rs.50000–55000	15

Qualitative Classification (Attribute)

This type of classification is based on descriptive characteristic or quality of data and is in accordance with non-measurable terms, like occupation, employment, religion, caste, literacy etc. (Fig. 1.1). If one group possesses a particular attribute, the other group will possess the opposite or other attribute. For example, if one group of people is literate, the other group will be illiterate. Similarly, if one group of people is honest, the other group will be dishonest.

Quantitative Classification (Numerical)

Quantitative characteristic of data (variable) which can be measured and expressed in numerical forms is the main basis of this type of classification. For example, monthly income of a group of people can be measured numerically, such as Rs. 12,000, Rs. 16,000, Rs. 20,000, Rs. 30,000 etc. (Table 1.8). Similarly, the monthly expenditure of people; height, weight and age of people can also be measured in numeric terms.

1.7.2 Tabulation of Data

Tabulation is the orderly and systematic arrangement of numerical data presented in columns and rows in order to extract information. It summarizes the data in a logical and orderly manner for the reasons of presentation, comparison and interpretation and makes the data brief and concise as they contain only the relevant figures. Gregory and Ward (1967) mentioned that 'Tabulation is the process of condensing classified data in the form of a table, so that it may be more easily understood and so that any comparisons involved may be more readily made'. The main aim of tabulation of data is to put the whole data set in a concise and logical manner. Connor (1937) stated that 'Table involves the orderly and systematic presentation of numerical data in a form designed to elucidate the problem under consideration'.

1.7.2.1 Essentials of an Ideal Table

No ideal method is there in the tabulation of data. Skill of data tabulation is generally a function of years of experience of the researcher. Nevertheless, the researcher should follow the following rules while tabulating the statistical information (Fig. 1.2):

1. **Table number:** When many tables are used, then they should be numbered like Tables 1 and 2 etc. for future reference. In case of several columns (more than four), they should also be numbered serially (Das 2009).
2. **Title of table:** Each table must have a clear and concise title which will convey the contents of the table.
3. **Stub:** Stub is the left-most column of the table which is clear and self-explanatory and used for representing the items and their headings. It is generally marked with rows in which an item is mentioned.
4. **Caption:** It is the title for columns other than the stub consisting of the upper part of the table.

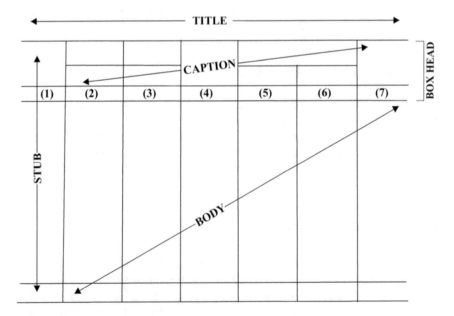

Fig. 1.2 Different parts of an ideal table

5. **Body of the table:** It is the main part of the table containing the clear and distinctive figures and data which are displayed in the table.
6. **Unit of measurement:** Units of measurement like Kg. for weight, ft. for height, Rs. for price etc. must be clearly mentioned in the column headings.
7. **Simplicity:** The table must be clear and simple keeping a balance between length and breadth in which the figures or values should be shown distinctly. The main columns and sub-columns should be indicated by heavy lines and light lines, respectively. The important figures in the table should be indicated by putting them in prominent place or in bold type.
8. **Arrangement:** The arrangement of data in the table depends on the nature of data, type of the table and the purposes for which they are intended. Data should be arranged in a logical sequence in the table. For example, the time series data must be arranged chronologically.
9. **Comparability:** The data should be arranged in the table in such a way that they become easy to compare. Comparable columns of figures should be kept as close as possible. In case of percentage figures, the basis of calculation of percentage should be mentioned near the figures to which they relate. Large number of figures should be rounded and indicated in thousands, millions etc.
10. **Source:** If the data have been collected and compiled from other sources, the source must be mentioned clearly in the foot-note.
11. **Total:** The total numbers of column must be mentioned at the bottom of the table and the row totals, if useful should also be mentioned.
12. **Foot-note:** The specific explanation about the figures should be mentioned as foot-note by using symbol (*), numbers (1, 2, 3,....) or English small letters (a, b, c,...).

Example A blank table has been prepared (Table 1.9) to show the discharge of water in pre-monsoon, monsoon and post-monsoon seasons by dividing into respective months during high and low tide at the places of Kolaghat, Soyadighi, Anantapur, Pyratungi, Dhanipur and Geonkhali on Rupnarayan River.

1.7.2.2 Types of Table

On the basis of purpose and uses, tables are of two types:

General Purpose Table

General purpose table, also called reference table, is generally voluminous in size and used as a repository of information. Special care and attention is required for the preparation of such table and the tabulation of information in it, because the information needed for reference may be obtained readily without any loss of time and effort (Bose 1980). These tables are prepared frequently by the concerned authority.

Table 1.9 Blank table to show season-wise water discharge in Rupnarayan River

Season	Month	Water discharge (m³/sec)															
		Kolaghat		Soyadighi		Anantapur		Pyratungi		Dhanipur		Geonkhali	
		High tide	Low tide	High tide	Low tide	High tide	Low tide	High tide	Low tide	High tide	Low tide	High tide	Low tide
Pre-monsoon	February												
	March												
	April												
	May												
	Average												
Monsoon	June												
	July												
	August												
	September												
	Average												
Post-monsoon	October												
	November												
	December												
	January												
	Average												

Name of the country	Total population (million)
China	1360
India	1210
USA	304
Indonesia	229
Brazil	193
Pakistan	165

Table 1.10 Simple table (Population size of some selected countries, 2011)

Source Human Development Report 2011, Oxford University Press, New Delhi.

Example The reports in tabular form prepared by different governmental and government-aided offices are the examples of general purpose table.

Special Purpose Table

Special purpose table or text table or summary table contains the summary of information and is used for special purposes. Generally, the table is small in size and prepared from the information gathered in the reference table. These tables are prepared suddenly.

Example Table prepared with the data collected about the smokers in India is a special purpose table.

Again, on the basis of nature and characteristics of classification of data, tables are of two types:

Simple Table

A simple table contains the data representing one characteristic only; information relating to other characteristics is left out (Table 1.10).

Complex Table

Complex table contains the data representing several characteristics. It shows the figures corresponding to a number of items (Table 1.11).

Table 1.11 Complex table (Hypothetical state of the Earth's atmosphere)

Altitude (m)	Hypothetical state of the Earth's atmosphere		
	Pressure (MPa)	Temperature (°C)	Density (kg/m^3)
0	0.1013	15.0	1.225
1000	0.898	8.5	1.1117
2000	0.759	2.0	1.0581
3000	0.701	−4.5	0.9093

1.7.3 Frequency Distribution

It is the method by which all the observations of a series are divided into a number of classes or groups, and the corresponding number of observations under each class are shown against its respective class. The number of times each value occurs is called *frequency,* and the table in which the distribution of observations against each variable or class of variables is shown is known as *frequency distribution table* or *frequency table.* A frequency distribution table contains a condensed summary of the original data. It represents not only the range of the values of a data series, but also shows the nature of their distribution throughout the range of the series.

The summarization of data into frequency distribution table entails much loss of details of the data. But it is very helpful and is an effective way for the treatment and interpretation of large volume of data. Some values, like different central values, values of dispersion and variability etc. of the data may be calculated easily from the frequency distribution table. The raw data (unorganized data having no form and structure) have no value in statistical analysis and interpretation. Array (arranged data in order of magnitude either in descending or in ascending order) has little significance in statistical analysis and interpretation. Frequency distribution of a large volume of information is very useful and significant in statistical analysis and interpretation.

On the basis of their nature, frequency distributions are of two types:

Simple (ungrouped) frequency distribution

In this type, the observations are not divided into groups or classes, the values of variables are shown individually (Table 1.12).

Grouped frequency distribution

In this type, the observations are divided into different classes or groups and the number observations in each class are shown as frequency (Table 1.13).

1.7.3.1 Important Terminologies Associated with Grouped Frequency Distribution

In grouped frequency distribution, the following terms are very useful and significant:

(a)   Class or class interval

Table 1.12 Simple
frequency distribution

Amount of rainfall (mm)	Frequency (Number of rainy days)
60	12
61	10
62	8
63	15
64	11
65	7
66	5

Table 1.13 Grouped
frequency distribution

Class interval (Temperature in °C)	Frequency (Number of days)
11–15	37
16–20	31
21–25	43
26–30	19
31–35	9
36–40	6
41–45	5

(b) Class limit (lower class limit and upper class limit)
(c) Class boundary (lower class boundary and upper class boundary)
(d) Class frequency (f_i) and Total frequency (N)
(e) Class mark or mid-value or mid-point of class interval (x_i)
(f) Class width or size of class interval (w_i)
(g) Frequency density (f_{di})
(h) Relative frequency (Rf_i)
(i) Percentage frequency

(a) **Class or class interval:** Large number of observations having wide range is usually classified into several groups according to the size of values. These groups are called *class interval* or simply classes. In Table 1.14, column (1), the class interval of temperatures (in °C) are 11–15, 16–20 etc. There are seven classes in the frequency distribution, the last class being 41–45.

Two ends of the classes are defined by class limits or boundaries. When two ends of a class are clearly specified, then it is called closed-end class but the class, in which one end is not clearly specified, is called an open-end class. When relatively few observations are far apart from the rest, then the construction of open-ended classes is required. Classes having no or zero frequency are called empty classes.

(b) **Class limit (Lower class limit and Upper class limit):** In case of grouped frequency distribution, the classes or class intervals, specified by pairs of values are arranged in such a way that the upper end (upper value) of one class does

Table 1.14 Frequency distribution table (Based on the data from Table 1.13)

Class Interval [1]	Class Frequency (f_i) [2]	Class Limit Lower [3]	Class Limit Upper [4]	Class Boundary Lower [5]	Class Boundary Upper [6]	Class Mark (x_i) [7]	Width of Class (w_i) [8]	Frequency Density (f_{di}) [9]	Relative frequency (Rf_i) [10]	Percentage frequency [11]
11–15	37	11	15	10.5	15.5	13	5	7.4	0.247	24.7
16–20	31	16	20	15.5	20.5	18	5	6.2	0.207	20.7
21–25	43	21	25	20.5	25.5	23	5	8.6	0.286	28.6
26–30	19	26	30	25.5	30.5	28	5	3.8	0.127	12.7
31–35	9	31	35	30.5	35.5	33	5	1.8	0.06	6
36–40	6	36	40	35.5	40.5	38	5	1.2	0.04	4
41–45	5	41	45	40.5	45.5	43	5	1	0.033	3.3
Total	$N = \sum f_i$ 150								1.00	100

not coincide with the lower end (lower value) of the immediately following class. These two extreme values, used to specify the limits of a class for the purpose of tallying the original observations into different classes are known as 'Class Limits'. The smaller value (lower end value) of the pair is called lower class limit, whereas the larger value (upper end value) is called upper class limit of a particular class (Sarkar 2015). In Table 1.14, the values 11, 16, 21, 26, 31, 36 and 41 (in column 3) are lower class limits, while the values 15, 20, 25, 30, 35, 40 and 45 (in column 4) are the upper class limits.

(c) **Class boundary (Lower class boundary and Upper class boundary):** Class boundaries are the limits up to which the two limits of each class may be extended to fill up the gap which exists between classes (Bose 1980). The upper boundary of one class coincides with the lower boundary of the immediately following class. The lower extreme value of the two boundaries is called the *lower class boundary* and the upper extreme value of the same is called the *upper class boundary* (columns 5 and 6 in Table 1.14).

Class boundary is calculated from class limit using the following formula:

$$Lower\ Class\ Boundary = \left[Lower\ Class\ Limit - \left(\frac{d}{2}\right)\right] \qquad (1.1)$$

$$Upper\ Class\ Boundary = \left[Upper\ Class\ Limit + \left(\frac{d}{2}\right)\right] \qquad (1.2)$$

where 'd' is the common difference between the upper class limit of any class (class interval) and the lower class limit of the next class (class interval). The observations are recorded to the nearest unit, $d = 1$ or the nearest tenth of a unit, $d = 0.1$ etc.

(d) **Class frequency and Total frequency:** *Class frequency* or simply *Frequency* is the number of observations (values) lying within a class. It is denoted by f_i. *Total frequency (N)* is the sum of all the class frequencies in a distribution. In other words, if all the class frequencies in a distribution are summed up, it indicates the *total frequency*. Total frequency shows the total number of observations considered in the frequency distribution. In Table 1.14, the class frequencies are 37, 31, 43,…5 (column 2) and the total frequency is 150. The working formula of total frequency (N) is as follows:

$$N = \sum_{i=1}^{n} f_i \qquad (1.3)$$

where n = number of class
f_i = frequency of the *i*th class.

(e) **Class mark or mid-value or mid-point of class interval:** The value lying exactly at the middle of a class interval is called *class mark* or *mid-value* (13, 18, 23 …43 are the class mark in column 7 in Table 1.14). The working formula for class mark is as follows:

$$Class\ Mark\ (x_i) = \frac{Lower\ Class\ Limit + Upper\ Class\ Limit}{2} \tag{1.4}$$

$$Or,\ Class\ Mark\ (x_i) = \frac{Lower\ Class\ Boundary + Upper\ Class\ Boundary}{2}$$
$$\tag{1.5}$$

The mid-value of the class or the class mark is considered as the representative value of the class for the computation of descriptive statistics like mean, mean deviation, standard deviation etc.

(f) **Class width or size of class interval:** The difference between the lower and upper class boundaries (but not class limits) is called *class width* or size of *class interval*. In Table 1.14, 5 is the class width (column 8) of this frequency distribution.

$$Width\ of\ class\ (w_i) = [Upper\ class\ boundary - Lower\ class\ boundary] \tag{1.6}$$

Generally, in a frequency distribution, equal width of the classes is preferred as it simplifies the calculation of some statistical measures (mean, median, mode, mean deviation, standard deviation etc.) in short-cut method. But in few cases, classes of unequal size may also be constructed when the values are highly dispersed in nature and some of them are few and far away from the rest. In such cases, the use of equal width may result in some 'Empty classes', i.e. classes with zero frequency.

(g) **Frequency density:** Number of frequency per unit class width is called the *frequency density* of a class. More the number of frequency per unit class width, more the *frequency density* and vice versa. The degree of concentration of frequency in a particular class is represented by the *frequency density* and is calculated by the following formula:

$$Frequency\ density\ (f_{di}) = \frac{Class\ Frequency}{Class\ width} = \frac{f_i}{w_i} \tag{1.7}$$

In Table 1.14, the frequency densities of different classes are 7.4, 6.2, 8.6 etc. (Column 9). Frequency density is used for the drawing of histogram in case of frequency distribution having unequal class width.

(h) **Relative frequency:** In a frequency distribution, the ratio between frequency of a particular class (f_i) and the total frequency (N) of the distribution is called *relative frequency* (Rf_i). The sum of all relative frequency in a distribution is equal to unity (1).

$$Relative\ Frequency\ (Rf_i) = \left(\frac{f_i}{N}\right) \tag{1.8}$$

and

$$\sum_{i=1}^{n} Rf_i = 1 \qquad (1.9)$$

where n = number of classes.

In Table 1.14, the relative frequencies of different classes are 0.247, 0.207, 0.286 etc. (Column 10).

(i) **Percentage frequency:** *Percentage frequency* is the class frequency when expressed as a percentage of the total frequency. In other words, when relative frequency is expressed in terms of percentage, then it is called as percentage frequency.

$$Percentage\ frequency = \frac{Class\ frequency}{Total\ frequency} \times 100 \qquad (1.10)$$

In Table 1.14, the percentage frequencies of different classes are 24.7, 20.7, 28.6 etc. (Column 11). The sum of all percentage frequencies in a distribution is equal to hundred percentage (100%).

1.7.3.2   Construction of Frequency Distribution Table

In grouped frequency distribution, the main questions are: (a) selection of the number of classes, (b) selection of class width and (c) selection of class limits and boundaries.

(a) **Selection of the number of classes:** There is no hard and fast rule in selecting the number of classes into which the observations would be divided. Generally, it depends on the nature of the data, number of observations in the series and the purpose for which the data are intended. It is generally agreed that the number of classes should neither be very large (to avoid lengthy and unwieldy frequency distribution) nor very small (information will be lost and the true pattern of the distribution of observations will be obscured). Normally, the number of classes should lie between 5 and 15, depending on the number of observations available. In case of small number of observations, some authors suggest the use of Sturges' formula:

$$n = 1 + 3.3\,logN \qquad (1.11)$$

where, n is the number of classes and N is the total number of observations in the data series.

(b) **Selection of class width:** Selection of the class width depends on the number of observations in the data series and the number of classes into which the observations are divided. For this purpose, at first we have to calculate the range (difference between highest and lowest value of the observations) of the data. If we like to have classes of equal width, then the width of the classes can be obtained by the formula:

$$Class\ width = \frac{Range(Highest\ value\ of\ the\ series - Lowest\ value\ of\ the\ series)}{Number\ of\ classes} \qquad (1.12)$$

$$Or,\ Class\ width = \frac{Range(Highest\ value\ of\ the\ series - Lowest\ value\ of\ the\ series)}{1 + 3.3\log N} \qquad (1.13)$$

Similarly, if the class width is known, the number of classes of the frequency distribution can be calculated by:

$$Number\ of\ classes = \frac{Range(Highest\ value\ of\ the\ series - Lowest\ value\ of\ the\ series)}{Class\ width} \qquad (1.14)$$

Example If the maximum and minimum values in a data series are 865 and 105, respectively, then the range of data will be $887-105 = 782$. In case of 8 number of classes, the width of classes (w_i) will be $\frac{782}{8} = 97.75$. It is very important to note that if the range of the data set is approximated to its nearest round figure which can easily be divided by the number of classes (n), the width of the class becomes easily recognizable and more comprehensive. In case of the above example, maximum value of 887 can be considered as 900 and the minimum value of 105 can be considered as 100. Thus, the range of the data becomes $900-100 = 800$ and the class width for 8 numbers of classes will be $\frac{900-100}{8} = \frac{800}{8} = 100$. So, the class width of 97.75 can easily be modified to 100 for practical applications.

Consideration of class width is very significant because it is an important determinant for the selection of class limits, class boundaries and class mark which are essentially used not only in preparing the frequency distribution table but also in the computations of different descriptive statistical measures (Sarkar 2015).

(c)   **Selection of class limits and boundaries:** Selection of class limit is made in two ways.

(i)   Exclusive method
(ii) Inclusive method

(i) Exclusive method: In this method, the upper limit of one class coincides with the lower limit of the next class, i.e. the upper limit of one class and the lower limit of the following class have the same figure and the same value. For example, 25–35, 35–45, 45–55 etc. (Table 1.15).

In this situation, the problem arises on account that in which class a value identical to the coinciding limits would be included. The problem is solved by excluding the identical value from the previous class and including it in the following class. In this sense, the upper limit of each class is considered as less than that limit while the lower limit of each class represents the exact value. Then, the class intervals of the above example will be stated as 25 to less than 35, 35 to less than 45, 45 to less than 55 etc.

Table 1.15 Exclusive and inclusive methods of selection of class limit

Exclusive method		Inclusive method	
Class limit (Weight in kg)	Frequency (Number of persons)	Class limit (Number of workers)	Frequency (Number of factory)
25–35	6	25–34	5
35–45	4	35–44	5
45–55	8	45–54	7
55–65	3	55–64	3
65–75	4	65–74	5

(ii) Inclusive method: In this method, the lower limit and upper limit of a particular class are included within the same class. Thus, the upper limit of one class does not coincide with the lower limit of the following class. Due to this, a gap exists between the upper limit of one class and the lower limit of the following class. For example, 25–34, 35–44, 45–54 etc. (Table 1.15). This method of classification may be applied for the grouped frequency distribution of discrete variables, such as number of family members, number of households, number of industrial workers etc., which can occur in integral values only. This method is not suitable to use in variables with fractional values like temperature, weight, height etc.

Thus, the nature and characteristics of the variable (continuous or discrete) under observation is important to decide whether the exclusive method or the inclusive method should be used for the selection of class limits. Exclusive method must be used for the classification of continuous variables whereas the inclusive method is suitable in case of discrete variables.

A frequency distribution table is drawn with (n + 1) rows and 6 columns (for equal class width) or 7 columns (for unequal class width). The column heads, from left to right are: class limits, class boundaries, class mark (x_i), class width (w_i), tally marks and frequency (f_i). In case of unequal classes, an additional column with headings of frequency density $(\frac{f_i}{w_i})$ is drawn.

Example-1 Heights (in metre) of 35 places from mean sea level are given below. Prepare a frequency distribution table from the given data.

412 350 307 308 432 342 357 297 328 375 356 429 329 240.
353 403 355 404 350 335 304 332 281 335 361 266 324 302.
406 366 337 345 343 227 364.

Solution:

Number of classes (n) = 1 + 3.3 log N.
= 1 + 3.3 log 35 [N = number of observations].
= 6.1294.
= 6 (nearest round figure).

Class width (w) = $\frac{\text{Range (Highest value of the series} - \text{Lowest value of the series)}}{\text{Number of classes}}$
= $\frac{432\,m - 227\,m}{6}$
= 34.17 m.
= 35 m.

Table 1.16 Frequency distribution table showing the height (in metre) from mean sea level

Class limit (Height in m)	Class Boundary (Height in m)	Class mark (x_i)	Class width (w_i)	Tally marks	Frequency (f_i)
227–261	226.5–261.5	244	35	||	2
262–296	261.5–296.5	279	35	||	2
297–331	296.5–331.5	314	35	ﷻﻟ |||	8
332–366	331.5–366.5	349	35	ﷻﻟ ﷻﻟ ﷻﻟ |	16
367–401	366.5–401.5	384	35	|	1
402–436	401.5–436.5	419	35	ﷻﻟ |	6
Total					$N = \sum f_i = 35$

Example-2 Mean monthly temperature (in °F) of 40 places are given below. Prepare a frequency distribution table from the given data.

29.4 49.5 39.6 45.7 53.8 39.7 36.6 58.7 34.4 39.7
54.4 54.7 51.5 62.5 23.0 80.8 30.3 27.7 44.0 35.7
56.1 60.2 72.2 50.8 33.4 56.3 32.4 59.2 48.2 45.1
42.7 52.1 24.2 68.6 66.0 39.5 43.4 36.4 44.1 56.6

Solution:

Number of classes (n) $= 1 + 3.3 \log N$.
$= 1 + 3.3 \log 40$ [N = number of observations].
$= 6.32$.
$= 6$ (nearest round figure).

$$\text{Class width(w)} = \frac{\text{Range (Highest value of the series} - \text{Lowest value of the series)}}{\text{Number of classes}}$$

$$= \frac{80.8\,^0\text{F} - 23.0\,^0\text{F}}{6}$$

$$= 9.63\,^0\text{F}$$

$$= 10\,^0\text{F}$$

Table 1.17 Frequency distribution table showing the mean monthly temperature (°F)

Class limit (Temperature in °F)	Class Boundary (Temperature in °F)	Class mark (x_i)	Class width (w_i)	Tally marks	Frequency (f_i)
23.0–32.9	22.95–32.95	27.95	10	ᴺᴴᴸ I	6
33.0–42.9	32.95–42.95	37.95	10	ᴺᴴᴸ ᴺᴴᴸ	10
43.0–52.9	42.95–52.95	47.95	10	ᴺᴴᴸ ᴺᴴᴸ	10
53.0–62.9	52.95–62.95	57.95	10	ᴺᴴᴸ ᴺᴴᴸ	10
63.0–72.9	62.95–72.95	67.95	10	|||	3
73.0–82.9	72.95–82.95	77.95	10	|	1
Total					$N = \sum f_i = 40$

1.7.3.3 Cumulative Frequency Distribution

The accumulated frequency upto or above some value of the variable is known as 'Cumulative frequency'. Cumulative frequency corresponding to a particular value of the variable can be defined as the number of observations smaller than or greater than that value (Das 2009). A cumulative frequency distribution is a form of frequency distribution in which the cumulative frequency upto each class is shown against the same class (Bose 1980). Cumulative frequency of any class is calculated by adding the frequency of each class to the total frequency of the previous classes. It represents the progressive total of the frequencies falling under each class.

A cumulative frequency distribution can be formed in two ways: (i) by less than method and (ii) by more than method. The number of observations 'upto' a given value is called less than cumulative frequency and the number of observations 'greater than' a value is called more than cumulative frequency. In the less than method, the frequencies are accumulated from the lowest class to upwards, but in more than method, the frequencies are accumulated from the highest class to downwards.

1.7.3.4 Uses of Cumulative Frequency Distribution

Cumulative frequency distribution is very significant and useful to determine the number of observations less than or greater than a particular value. It is very helpful in finding (a) the number of observations less than or below any given value (b) the number of observations more than or above any given value (c) the number of observations falling between two specific values.

Table 1.18 Cumulative frequency distribution table using the data of Table 1.16

Class limit (Height in m)	Class Boundary (Height in m)	Frequency (f_i)	Cumulative Frequency (F)			
			Less than	F	More than	F
227–261	226.5–261.5	2	226.5	0	226.5	35 (0 + 6 + 1 + 16 + 8 + 2 + 2)
262–296	261.5–296.5	2	261.5	2 (0 + 2)	261.5	33 (0 + 6 + 1 + 16 + 8 + 2)
297–331	296.5–331.5	8	296.5	4 (0 + 2 + 2)	296.5	31 (0 + 6 + 1 + 16 + 8)
332–366	331.5–366.5	16	331.5	12 (0 + 2 + 2 + 8)	331.5	23 (0 + 6 + 1 + 16)
367–401	366.5–401.5	1	366.5	28 (0 + 2 + 2 + 8 + 16)	366.5	7 (0 + 6 + 1)
402–436	401.5–436.5	6	401.5	29 (0 + 2 + 2 + 8 + 16 + 1)	401.5	6 (0 + 6)
		$N = \sum f_i = 35$	436.5	35 (0 + 2 + 2 + 8 + 16 + 1 + 6)	436.5	0

Cumulative frequency may be represented in relative or percentage form. When it is represented in percentage, it is known as cumulative percentage. It is very helpful for the comparison between frequencies.

Example-1

See Table 1.18.

Example-2

See Table 1.19.

1.8 Methods of Presentation of Geographical Data

Presentation of data means the demonstration of the data in an attractive and lucid manner to make them easily understandable to all. Suitable and accurate visualization of the collected data becomes helpful for their proper understanding, analysis and explanation. Geographical data can be represented and portrayed in the following four ways:

Table 1.19 Cumulative frequency distribution table using the data of Table 1.17

Class limit (Temperature in °F)	Class Boundary (Temperature in °F)	Frequency (f_i)	Cumulative Frequency (F)			
			Less than	F	More than	F
23.0–32.9	22.95–32.95	6	22.95	0	22.95	40 (0 + 1 + 3 + 10 + 10 + 10 + 6)
33.0–42.9	32.95–42.95	10	32.95	6 (0 + 6)	32.95	34 (0 + 1 + 3 + 10 + 10 + 10)
43.0–52.9	42.95–52.95	10	42.95	16 (0 + 6 + 10)	42.95	24 (0 + 1 + 3 + 10 + 10)
53.0–62.9	52.95–62.95	10	52.95	26 (0 + 6 + 10 + 10)	52.95	14 (0 + 1 + 3 + 10)
63.0–72.9	62.95–72.95	3	62.95	36 (0 + 6 + 10 + 10 + 10)	62.95	4 (0 + 1 + 3)
73.0–82.9	72.95–82.95	1	72.95	39 (0 + 6 + 10 + 10 + 10 + 3)	72.95	1 (0 + 1)
		$N = \sum f_i = 40$	82.95	40 (0 + 6 + 10 + 10 + 10 + 3 + 1)	82.95	0

1.8.1 Textual Form

Textual presentation is the most raw and vague form of representation of geographical data. In textual form, data are presented in paragraph or in sentences. When the amount of data is not too large, then this form of presentation is more appropriate and effective. In textual presentation, mainly the important characteristics are enumerated giving emphasis on the most significant figures and highlighting the most striking attributes of the data set. Significant figures and attributes may be the summary statistics like maximum and minimum value, mean, median, mean deviation, standard deviation etc.

Example Out of 180 sediment samples studied in Rupnarayan River, approximately, 63.80% of the sediments are very fine sand, 14.76% are fine sand and 21.44% are coarse silt type. In dry season, more than 60% sediments are moderately to well sorted but in monsoon season 63.85% sediments are poorly to very poorly sorted.

Around 55% of the sediments are of fine and very fine skewed type, 33% of samples are near symmetrical and remaining 12% are of coarse skewed type.

1.8.1.1   Advantages and Disadvantages of Textual Form

Advantages

1.   Easy to understand.
2.   It enables one to give emphasis on certain important features of the data presented.

  Disadvantages

1.   One has to go through the complete reading of the text for comprehension.
2.   Boring to read especially if too lengthy.
3.   Reader may skip the statements.

1.8.2   Tabular Form

Tabular presentation of geographical data is very important and easily understandable to all. It is one of the most commonly used forms of representation of data as tables are very easy to construct and understand. A table makes possible representation of even large amounts of data in a lucid, attractive and organized manner. Tabulation is the orderly and systematic arrangement of numerical data presented in columns and rows in order to extract information. It summarizes the data in a logical and orderly manner for the reasons of presentation, comparison and interpretation and makes the data brief and concise as they contain only the relevant figures (Table 1.20) [Detailed discussion in Sect. 1.7.2].

Table 1.20 Tabular presentation of data (% of sand, silt and clay in bed sediments of Rupnarayan River)

Locations	Sand-silt-mud proportion (%)								
	Pre-monsoon season			Monsoon season			Post-monsoon season		
	Sand	Silt	Clay	Sand	Silt	Clay	Sand	Silt	Clay
Kolaghat	68–91	8–30	1–15	76–91	8–17	1–12	71–86	9–20	5–20
Soyadighi	60–78	12–25	8–18	70–86	8–21	2–18	70–84	7–21	8–18
Anantapur	45–78	13–48	6–38	59–86	10–42	4–17	55–78	14–39	4–22
Pyratungi	54–79	8–32	12–26	73–87	9–20	4–15	56–85	4–17	9–18
Dhanipur	38–76	11–61	1–41	52–84	10–45	3–18	45–75	24–54	1–25
Geonkhali	46–78	9–40	12–38	61–87	9–18	4–19	49–74	10–32	16–21

Source Field survey and laboratory experiment.

1.8.2.1   Advantages and Disadvantages of Data Representation in Table

Advantages

The advantages of tabulation of data are as follows:

1. By tabulation, data are arranged systematically and logically in concise form.
2. Tabulation enables the data to be easily understandable and it is more impressive than textual presentation.
3. It is very useful to detect the errors and exclusions in the data.
4. Recurrence of explanatory terms and phrases can be avoided.
5. The nature and characteristics of data can easily be understood at a glance in tabular form.
6. Comparison and interpretation of statistical data becomes easy.

Disadvantages

1. Tabular presentation does not give a detailed view of the data, unlike textual (descriptive) presentation.
2. It is only helpful to identify the differences of points or if we want to tally two or more things.

1.8.3   Semi-Tabular Form

It is the combination of textual and tabular form of data presentation. This is also called partial-tabular presentation of data. It is helpful for the easy comparison because the numerical figures are separately presented from the text.

Example Overall literacy rates in different census years after independence in India are:

- 16.67% in 1951
- 24.02% in 1961
- 29.45% in 1971
- 36.23% in 1981
- 42.84% in 1991
- 54.51% in 2001
- 64.32% in 2011.

1.8.4   Graphical Form (Graphs, Diagrams and Maps)

In addition to all the above mentioned methods, classified and tabulated geographical data can suitably and easily be represented through different *graphs* (line graph, climograph, Lorenz curve, rank-size graph, frequency graph etc.), *diagrams* (bar diagram, pie-diagram, rectangular diagram etc.) and *maps* (choropleth map, chorochromatic map, choroschematic map etc.). All the graphs, diagrams and maps are drawn following various geometric methods, thus it is known as geometric representation of data. Representation of geographical data by graphical, diagrammatic and mapping techniques is very popular, attractive and easy to understand to the geographers, researchers and to the common literate people also.

References

Bose A (1980) Statistics. Calcutta Book House, 1/1 Bankim Chatterjee Street, Calcutta 700073

Connor LR (1937) Statistics in theory & practice, 2nd edn, Sir Isaac Pitman & Sons, Inc

Das NG (2009) Statistical methods, vol. I & II. McGraw Hill Education (India) Pvt Ltd, ISBN: 978-0-07-008327-1

Galtung J (1968) A structural theory of integration. J Peace Res 5(4):375–395

Gregory H, Ward D (1967) Statistics for business studies. McGraw-Hill. ISBN:9780070944909

Kapur SK (1995) Elements of practical statistics. Oxford & IBH Publishing Co Pvt Ltd., New Delhi

Khan MAT (2006) Quantitative techniques in geography. Perfect Publications, Dhaka. ISBN: 984-8642-02-1

Pal SK (1998) Statistics for Geoscientists: Techniques and Applications. Concept Publishing Company, New Delhi. ISBN: 81-7022-712-1

Sarkar A (2015) Practical geography: a systematic approach. Orient Blackswan Private Limited, Hyderabad, Telengana, India. ISBN: 978-81-250-5903-5

Young PV (1994) Scientific social surveys and research. Prentice Hall of India Private Limited, New Delhi

Chapter 2
Representation of Geographical Data Using Graphs

Abstract Suitable, accurate and lucid presentation and visualization of geographical data using various types of graphs become helpful for their correct analysis, explanation and realization for proper understanding of the real world. It is very simple, attractive and easily recognizable not only to the geographers or efficient academicians but also to the common literate people. This chapter includes a detailed classification of all types of graphs and the discussion of various types of co-ordinate systems with illustrations as an essential basis of the construction of graphs. Different types of bi-axial (arithmetic and logarithmic graph, climograph etc.), tri-axial (ternary graph), multi-axial (spider graph, polar graph etc.) and special graphs (water budget graph, hydrograph, rating curve, Lorenz curve, rank-size graph, hypsometric curve etc.) have been discussed with suitable examples in terms of their suitable data structure, necessary numerical calculations, methods of construction, appropriate illustrations, and advantages and disadvantages of their use. Systematic and step-by-step discussion of methods of their construction helps the readers for easy and quick understanding of the graphs. The difference between arithmetic and logarithmic graphs is explained precisely with proper examples and illustrations. Different types of frequency distribution graphs have been explained with suitable data, necessary mathematical and statistical computations, and proper illustrations. All types of graphs represent a perfect co-relation between the theoretical knowledge of various geographical events and phenomena and their realistic implications with suitable examples.

Keywords Graphs · Co-ordinate system · Bi-axial graph · Arithmetic and logarithmic graph · Tri-axial graph · Multi-axial graph · Special graph · Frequency distribution graph

2.1 Concept of Graph

The method of representation of *geographical data* in *graphs* has been developed to avoid the difficulties arising from their tabular presentation and for their better understanding also. This technique is very helpful to understand and explain the relationship between various geographic data, to indicate the trends of different geographic variables and to make a comparison between them. Graph is the most familiar and

© The Author(s), under exclusive license to Springer Nature Singapore Pte Ltd. 2021
S. K. Maity, *Essential Graphical Techniques in Geography*, Advances in Geographical and Environmental Sciences, https://doi.org/10.1007/978-981-16-6585-1_2

conventional method in which a series of geographical data are represented following a suitable *co-ordinate system* on a reference frame.

2.2   Types of Co-ordinate System

The four main types of co-ordinate systems are:

(1)   Cartesian or rectangular co-ordinate system
(2)   Polar co-ordinate system
(3)   Cylindrical co-ordinate system
(4)   Spherical co-ordinate system

2.2.1   Cartesian or Rectangular Co-ordinate System

Cartesian co-ordinate or *rectangular co-ordinate system* is a co-ordinate system that identifies each point distinctively on a plane with the help of a set of numerical co-ordinates. Co-ordinates are the signed (either positive or negative) distances to the specific point from two perpendicular oriented lines. Both the co-ordinates and lines are measured and represented in the same unit of length. This co-ordinate system provides a technique of portraying graphs and representing the positions of points on a two-dimensional (2D) surface as well as in a three-dimensional (3D) space.

On a *two-dimensional (2D) surface,* while constructing a graph, at first a horizontal line XX' called *abscissa* and a vertical line YY' called *ordinate* (both are known as co-ordinate axes) are drawn which intersect each other at right angles and the whole plotting area is divided into four parts called *quadrants* (Fig. 2.1). The point of intersection of these two axes is called *point of origin* ('*O*') or zero-point having $x–y$ co-ordinate (0, 0). All the distances along these two axes are always measured from this zero point. The rightward and upward measurement of distances from the zero point indicates the positive values, whereas the leftward and downward measurements represent the negative values. The values of both '*x*' and '*y*' are positive in the first quadrant, but in the second quadrant, the values of '*x*' and '*y*' are negative and positive, respectively. In the third quadrant, both '*x*' and '*y*' are negative, while in the fourth quadrant, '*x*' value is positive but '*y*' value is negative (Fig. 2.1). Most of the geographical data collected from field investigations are positive in nature and hence these data are represented in the first quadrant in which the values of '*x*' and '*y*' both are positive. The use of second, third and fourth quadrants are comparatively less in statistical and geographical analysis.

In *Cartesian three-dimensional (3D) space,* another important axis oriented at right angles to the xy plane is added. This axis passes through the origin of

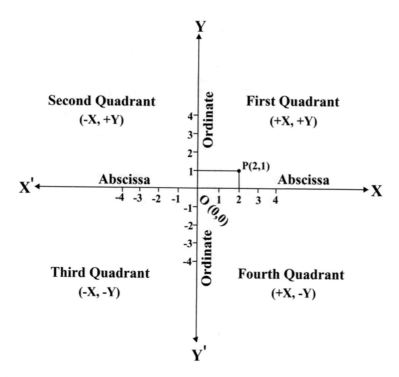

Fig. 2.1 Position of independent and dependent variables in different quadrants (Cartesian co-ordinate system)

the xy plane and is called the 'Z' axis, representing the height. Positions or co-ordinates of points ('P', 'Q', 'R', 'S' in Fig. 2.2) are determined based on the east–west (x), north–south (y) and up–down (z) displacements of points from the origin 'O' (0, 0, 0).

2.2.2 Polar Co-ordinate System

Polar co-ordinate system is another common and important co-ordinate system for the plane. This two-dimensional co-ordinate system specifies the location of each point on a plane by the measurement of the distance from a reference point (called pole) and an angle from a reference direction (called polar axis). The location of the point is obtained by measuring the signed distance from the origin (pole) and the given angle (measured counter-clockwise) from the polar axis. For a given distance from the origin 'r' and angle from polar axis 'θ', the co-ordinate of any point ('P' in Fig. 2.3) is (r, θ). The pole is characterized by $(0, \theta)$ for any value of θ.

Polar co-ordinate system is extended to three dimensions in two methods like cylindrical co-ordinate system and spherical co-ordinate system.

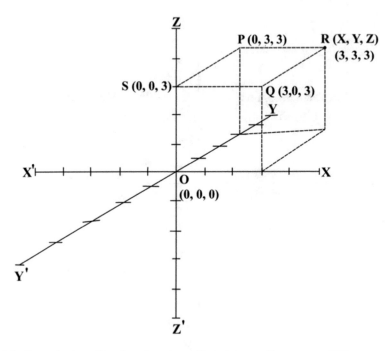

Fig. 2.2 Determination of location of a point on Cartesian co-ordinate system (3D)

Fig. 2.3 Determination of
location of a point on polar
co-ordinate system

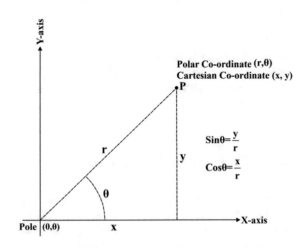

The relation between Cartesian co-ordinate (x, y) and polar co-ordinate (r, θ) is
that (Fig. 2.3)

$$\sin \theta = \frac{y}{r} \text{(sine function for } y) \tag{2.1}$$

where $y = r \sin \theta$

$$\text{and } \cos \theta = \frac{x}{r} \text{(cosine function for } x\text{)} \tag{2.2}$$

where $x = r \cos \theta$.
Again,

$$r = \sqrt{x^2 + y^2} \text{(Pythagoras theorem to find the long side, i} \cdot \text{e. the hypotenuse)} \tag{2.3}$$

and

$$\theta = \tan^{-1}\frac{y}{x} \text{(tangent function to find angle)} \tag{2.4}$$

Conversion from Cartesian to polar co-ordinate system
Example: What is (4, 3) in polar co-ordinates?

Solution
Using Pythagoras theorem (Eq. 2.3) we have

$$r = \sqrt{x^2 + y^2}$$

$$r = \sqrt{4^2 + 3^2}$$

$$r = \sqrt{16 + 9}$$

$$r = \sqrt{25}$$

$$r = 5$$

Using tangent function (Eq. 2.4) we have

$$\theta = \tan^{-1}\frac{y}{x}$$

$$\theta = \tan^{-1}\frac{3}{4}$$

$$\theta = \tan^{-1}0.75$$

$$\theta = 36.87°$$

Answer: The point (4, 3) is (5, 36.87°) in polar co-ordinates.

Conversion from polar to Cartesian co-ordinate system
Example: What is (5, 36.87°) in Cartesian co-ordinates?

Solution
Using the sine function (Eq. 2.1) we have

$$\sin\theta = \frac{y}{r}$$

$$\sin\theta = \frac{y}{5}$$

$$y = 5 \times \sin 36.87°$$

$$y = 5 \times 0.6$$

$$y = 3$$

Using the cosine function (Eq. 2.2) we have

$$\cos\theta = \frac{x}{r}$$

$$\cos\theta = \frac{x}{5}$$

$$x = 5 \times \cos 36.87°$$

$$x = 5 \times 0.8$$

$$x = 4$$

Answer: The point (5, 36.87°) is (4, 3) in Cartesian co-ordinates.

2.2.3 Cylindrical Co-ordinate System

Cylindrical co-ordinate system is the extension of the polar co-ordinates by adding the z-axis along with the height of a right circular cylinder. The z-axis in this co-ordinate system is the same as in Cartesian co-ordinate system (3D). The addition of z-axis in polar co-ordinate system gives a triple (r, θ, z) (Fig. 2.4). In some texts, ρ is used in place of r to denote the distance from the origin to the foot of the perpendicular to avoid confusion. In terms of Cartesian co-ordinate system

Fig. 2.4 Determination of
location of a point on
cylindrical co-ordinate
system

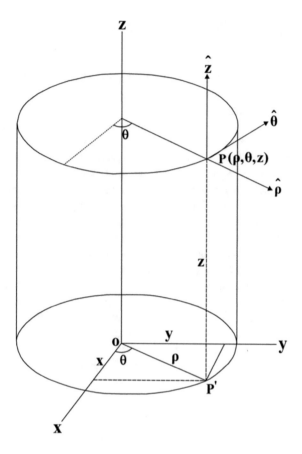

$$x = \rho \cos\theta \qquad (2.5)$$

$$y = \rho \sin\theta \qquad (2.6)$$

$$z = z \text{ (height)}$$

Again, in inverse relation these become

$$\rho = \sqrt{x^2 + y^2} \qquad (2.7)$$

$$\theta = \tan^{-1}\frac{y}{x} \qquad (2.8)$$

$$z = z \text{ (height)}$$

2.2.4 Spherical Co-ordinate System

Spherical coordinate (also known as spherical polar coordinate) system is a curvilinear coordinate system in which positions of points are defined on a sphere or spheroid. In spherical co-ordinate, the value of z co-ordinate is converted into ϕ giving a triple (r, θ, ϕ). Here, r is the distance of a point (say P in Fig. 2.5a, b) from the origin (radial distance) and θ (azimuthal angle) is the angle between the x-axis and the line joining the origin to P', the foot of the perpendicular from the point P (Fig. 2.5a, b) in the x–y plane. The angle θ is complementary to the longitude $[0 \leq \theta < 2\pi]$ and is denoted as λ when referred to as longitude. The angle ϕ (polar angle, zenith angle or colatitude) is the angle made by the radius vector (the vector which connects the point P with origin) with respect to the z-axis. It is complementary to the latitude $[0 \leq \phi \leq \pi]$ and is represented as $\phi = 90° - \delta$, where δ is the latitude. Conventionally, (r, θ, ϕ) is used in mathematics to represent radial distance, azimuthal angle and polar angle, respectively. Sometimes, especially in physics θ and ϕ are reversely used, i.e. θ indicates polar or zenith angle and ϕ indicates azimuthal angle. Then, (r, θ, ϕ) represent radial distance, polar angle and azimuthal angle, respectively. In spherical co-ordinate, the symbol ρ is frequently used instead of r to avoid the confusion with the value r in 2D polar coordinate systems, i.e. then (r, θ, ϕ) becomes (ρ, θ, ϕ).

The transformation from spherical co-ordinates (r, θ, ϕ) [radial, azimuthal, polar] to Cartesian co-ordinates (x, y, z) is given by

$$x = r \cos\theta \, \sin\phi \qquad (2.9)$$

$$y = r \sin\theta \, \sin\phi \qquad (2.10)$$

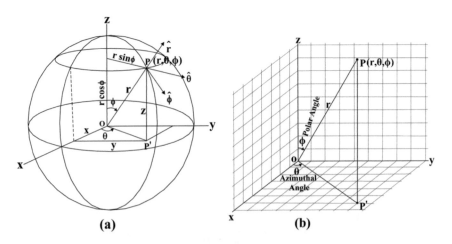

(a)   **(b)**

Fig. 2.5 Determination of location of a point on spherical co-ordinate system

$$z = r \cos\phi \qquad (2.11)$$

Again, the inverse relation indicates

$$r = \sqrt{x^2 + y^2 + z^2} \qquad (2.12)$$

$$\theta = \tan^{-1}\left(\frac{y}{x}\right) \qquad (2.13)$$

$$\phi = \tan^{-1}\left(\frac{\sqrt{x^2 + y^2}}{z}\right) = \cos^{-1}\left(\frac{z}{r}\right) \qquad (2.14)$$

2.3   Selection of Scale in Constructing a Graph

While constructing a graph, the selection of scale should be done carefully keeping two things in mind: (i) the nature and range of the entire data set and (ii) the size of the graph paper. Conventionally, the independent variable is shown along the x-axis (abscissa) while the dependent variable is shown along the y-axis (ordinate). It is not mandatory to make the scale of the x-axis and y-axis identical. In time series data, the scale of the x-axis starts from the lowest value of the given variable or the starting time of the time series, whereas the scale of the y-axis starts from the value zero (0). But in the case of frequency distribution, the scale along the y-axis starts from zero while the scale along the x-axis may start from zero or with the value one point before the lowest value of the measured variable (Saksena 1981). After selecting the suitable scale, all the points are plotted on the graph paper and then the obtained points are joined by straight lines but it is not mandatory. Though several types of graphs are there, the selection of suitable graph mainly depends on the type and nature of data and the objective of the study or research.

In geographical research, the collected, classified, tabulated and summarized data are represented graphically to make them easily understandable and comprehensive. As most of the graphical representation of geographical data is done by geometrical methods, thus it is also known as the geometrical representation of data.

2.4   Advantages and Disadvantages of the Use of Graphs

Graphical representation of various geographical data possesses some advantages as well as some disadvantages:

Advantages

1. Graphical representation is more attractive and appealing to the eyes, which leaves an enduring impression on the mind and is thus easily understandable to all.
2. The trend and tendency of the values of geographical variables (time series data) can be easily understood.
3. It is very effective and useful to understand the nature and characteristics of complex geographical data sets.
4. It helps in making a comparison of two or more sets of geographical data.
5. The relationship between several sets of geographical variables can be effectively shown by this method.
6. Any type of inaccuracy and error in geographical data becomes perceptible by their graphical representation.
7. It is useful for the interpolation of values of geographical variables.
8. Median, mode, quartile and other descriptive statistics can be easily calculated and estimated by graphical representation of geographical data.

Disadvantages

1. Overall and detailed information of geographical data cannot be obtained from their graphical representation.
2. It reveals only the approximate position; it seldom reflects the perfect values.
3. It is time-consuming to prepare the graph.
4. Selection of inappropriate graph may lead to erroneous conclusions and decisions.
5. A high degree of variability between the values of geographic data may obstruct the purpose of graphical representation.
6. Representation and understanding of several numbers of geographic variables become difficult in the graph.
7. Sometimes the graph shows the difficulty to understand the inefficient, illiterate and common people.

2.5 Types of Graphical Representation of Data

Graphical representation of data can be broadly classified into the following heads given in Table 2.1.

2.5.1 Bi-axial Graphs or Line Graphs or Historigram

The values of two geographical elements or variables are represented along the sets of 'X' and 'Y' axes on a reference frame. The line graphs are generally drawn to represent the time series data like temperature, rainfall, birth rates, death rates, growth of population etc.

Table 2.1 Types of graphs

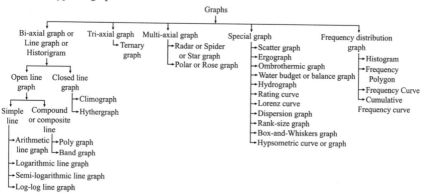

The values of different *geographical variables* change over time. A series of observations recorded in accordance with the time of occurrence is called time series data. The graphical representation of classified and summarized time series data is called historigram in which time is considered as independent variable and the corresponding geographical values are taken to be dependent variable. For the comparison of temporal changes of two or more variables expressed in the same unit of measurement, two or more historigrams are drawn. There are numerous geographical data which can effectively and successfully be represented by historigram.

Time series graph or historigram indicates two important facts of geographical data:

(1) Measurement and analysis of the changes of uni-variate geographical data.
(2) Comparison of changes of two or more geographical variables.

For the construction of historigram, time (year, month, day etc.) is shown along the 'X'-axis and the corresponding geographical variable (temperature, rainfall, number of landslide hazard, volume of water discharge in a river, number of population, volume of population migration, amount of agricultural or industrial production etc.) is shown along the 'Y'-axis following a suitable scale. Plotting of the values of any geographical phenomenon or event with respect to time provides some points which are then joined by a line called line graph or historigram (Figs. 2.6 and 2.7).

For example, the increase of population in Kolkata Urban Agglomeration (KUA) with the advancement of time (Table 2.2) can be represented by historigram. Here, different years are shown on the 'X'-axis and the total population are shown on the 'Y'-axis and then the plotted points are joined by a line (Fig. 2.6).

Similarly, the variation of rice production in different years (Table 2.3) can be represented by historigram. Here, different years are shown on the 'X'-axis and the amount of production of rice are shown on the 'Y'-axis and then the plotted points are joined by a line (Fig. 2.7).

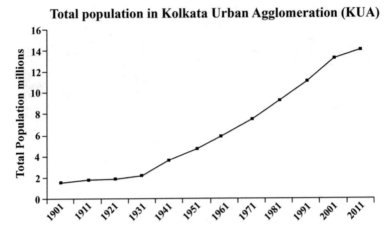

Fig. 2.6 Line graph (Historigram) showing the temporal changes of total population in Kolkata Urban Agglomeration (KUA) *Source* Census of India

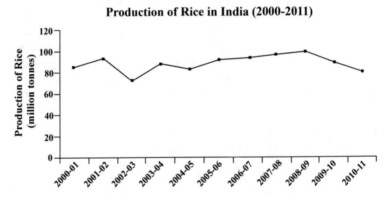

Fig. 2.7 Line graph or Historigram (Production of rice in India, 2000–2011) *Source* Directorate of Economics and Statistics (Government of India)

2.5.1.1 Open Line Graph

Simple Line Graph

When the line graph represents the values of only a single variable or element or fact, then it is called a *simple graph.*

Arithmetic Graph

Use of *arithmetic or linear scale* on the horizontal (X-axis) and vertical (Y-axis) axes to represent geographical data using line graph is more frequent and common.

Table 2.2 Data for line graph or historigram (Temporal change of total population in Kolkata Urban Agglomeration)	Year	Total population (in millions)
	1901	1.51
	1911	1.74
	1921	1.88
	1931	2.14
	1941	3.62
	1951	4.67
	1961	5.98
	1971	7.42
	1981	9.19
	1991	11.02
	2001	13.21
	2011	14.03

Table 2.3 Data for line graph or historigram (Production of rice in India, 2000–2011)	Year	Production of rice (million tons)
	2000–01	84.98
	2001–02	93.34
	2002–03	71.82
	2003–04	88.53
	2004–05	83.13
	2005–06	91.79
	2006–07	93.36
	2007–08	96.69
	2008–09	99.18
	2009–10[a]	89.13
	2010–11[b]	80.41

[a]Fourth advance estimates as released on 19.07.2010
[b]First advance estimates as released on 23.09.2010
Source Directorate of Economics and Statistics (Government of India)

On an arithmetic scale, equal amounts or values are represented by equal distances, i.e. the values of a data series plotted on an arithmetic scale increase or decrease at a constant rate (even spaces between numbers). Thus, the distance from a value of 1 to 2 (distance is 1) is equal to that of the distance from 2 to 3 (distance is 1), 3 to 4 (distance is 1), 4 to 5 (distance is 1) and so on (Fig. 2.8a). Representation of data on an arithmetic or a linear scale would produce a curving line, descending at a declining (getting lower) angle for a diminishing series of values and ascending at a rising (getting higher) angle for a growing series of values.

The major advantages and disadvantages of using arithmetic graphs are:

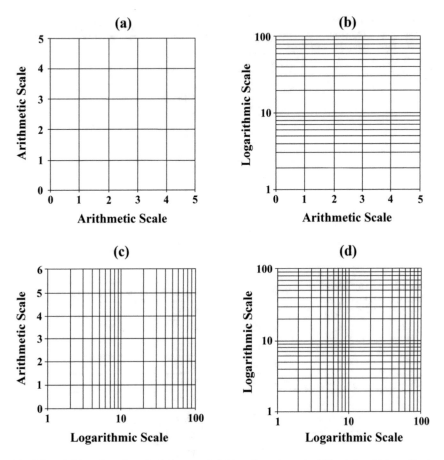

Fig. 2.8 **a** Arithmetic scale on both the axes, **b** Arithmetic scale on the 'X'-axis but the logarithmic scale on the 'Y'-axis, **c** Arithmetic scale on the 'Y'-axis but the logarithmic scale on the 'X'-axis, and **d** Logarithmic scale on both the axes

Advantages

(1) Presentation of geographical data with a line graph on arithmetic scales is very easy and simple because basic mathematical principles are applied.

(2) Arithmetic line graphs are very easy to read and understand. Most of the readers expect a level twice as high to be twice as large.

(3) Zeroes or negative values can be easily represented on arithmetic scales.

(4) These graphs are useful for the representation and understanding of the absolute changes of values of geographical variables.

Disadvantages

(1) These graphs only show the absolute changes of values; however, these do not show the relative changes. Thus, these are not useful for comparing the relative changes (percentage) of values of geographical variables.

Logarithmic Graph

Representation of *geographical data* with a line graph on a *logarithmic scale* (equal scale between powers of 10) is an alternative and more useful technique in comparing the rate of change of values. These graphs are more useful and effective to understand and compare the relative changes (percentage) of a set of values rather than their absolute amounts of changes. Log-graph is commonly used when the range of values of the variable is very large and an increase or decrease of the values occurs roughly at a constant ratio. On a logarithmic scale, equal distances stand for equal ratios. For instance, the distance from 1 to 2 is equal to that from 2 to 4 $\left(\frac{2}{4} = \frac{1}{2}\right)$, 4 to 8 $\left(\frac{4}{8} = \frac{1}{2}\right)$, 8 to 16 $\left(\frac{8}{16} = \frac{1}{2}\right)$ and so on at each interval, in the ratio 1:2 (vertical axis of Fig. 2.8b, horizontal axis of Fig. 2.8c and both axes of Fig. 2.8d). Representation of data on a logarithmic scale clearly depicts the percentage increase or decrease between two data values.

In log-graph paper, the axes (either X-axis or Y-axis or both) are divided into several parts of equal length, known as cycles. A single cycle corresponds to a tenfold increase of values of variables, and similarly, two cycles indicate the 100-fold increase of values. The value at the top of the first cycle is ten times more than that of the value at the bottom of it and the value at the top of the second cycle is ten times more than the value at the bottom of the second cycle (the top of the first cycle), i.e. hundred times more than that of the value at the bottom of the first cycle (vertical axis of Fig. 2.8b, horizontal axis of Fig. 2.8c and both axes of Fig. 2.8d). It is because of the principle that a common logarithm is a power to which 10 should be increased to generate a specified number. Thus, $100 = 10^2$, $1000 = 10^3$ and the logarithm of 100 and 1000 are 2 and 3, respectively, and so on. Log scale can never start with zeroes or negative values, as $\log(0) = \infty$ (infinity). So, any positive value should be taken at the origin by the user. Based on the selection of logarithmic scale (either on the X-axis or Y-axis or both the axes), log-graphs are of two types.

Semi-logarithmic Graph

A *semi-logarithmic or semi-log line graph* has one axis on a logarithmic scale (equal scale between powers of 10) and another axis on an arithmetic or linear scale (even spaces between numbers) (Fig. 2.8b, c). These graphs are useful for the data with exponential relationships, or where a single variable covers a large range of values. A set of geographical data plotted using a logarithmic scale on the y-axis will resemble a straight line, slanting up or down based on the nature of the data values. When the values increase or decrease at a constant rate, it will appear as a straight line.

Table 2.4 Database for arithmetic and logarithmic line graph (Age and sex-specific variation of death rates)

Age group	Number of deaths (per year)	
	Male	Female
<15	15	20
15–19	17	20
19–24	23	24
25–29	27	45
30–34	33	105
35–39	60	210
40–44	110	318
45–49	235	480
50–54	470	625
55–59	820	820
60–64	1340	1205
65–69	2110	1508
70–74	2905	1750
75–79	3380	1820
80–84	3385	2010
84+	2000	1325

Log–Log Graph

In *log–log graph,* both the X (horizontal) and Y (vertical) axes are represented using the logarithmic scale in which equal distances measure equal ratios (Fig. 2.8d).

The given two line graphs (Figs. 2.9 and 2.10) show the difference between the two scales when representing the same data, i.e. age-specific number of deaths of the male and female population (Table 2.4). The first line graph (Fig. 2.9) has been drawn using the arithmetic scale on both axes. The graph demonstrates that female death rates are slightly or little higher than the male death rates until about age group 50–54. In the age group 55–59, the male death rate has started to exceed the female death rate and in the age group 65–69, the rate becomes much higher and stays much higher. However, the second line graph (Fig. 2.10) has been drawn using a logarithmic scale on the 'y'-axis. In this graph, the female death rates for the younger age groups appear somewhat higher in comparison to the male death rates and the relative (percentage) differences in the death rates for the older age groups are not as higher as apparent in the arithmetic line graph.

So, the *logarithmic graph* highlights a possible significant difference between the death rates of the male and female population in the younger age groups, whereas this difference has been lost in the arithmetic line graph because of the plotting of the higher absolute values for the older age groups.

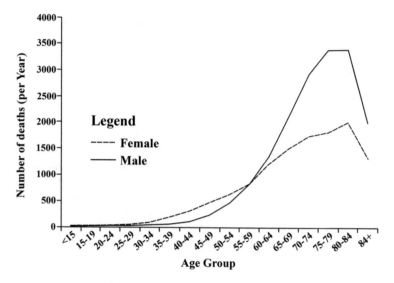

Fig. 2.9 Arithmetic graph (Number of male and female deaths per year)

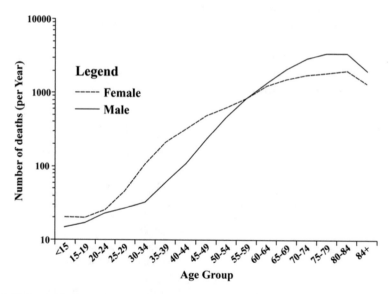

Fig. 2.10 Logarithmic graph (Number of male and female deaths per year)

Advantages and Disadvantages of Using Logarithmic Graph

Advantages

(1)   Logarithmic line graphs show the relative changes, i.e. these graphs are useful to identify and understand the relative changes (percentage) of values of geographical variables.

(2)   In terms of comparative study, the logarithmic line graphs provide a more complete depiction and explanation of the relationship that exists between sets of data.

Disadvantages

(1)   Representation of zeroes or negative values on logarithmic line graphs is not possible. Also, these graphs are not so easy to construct like arithmetic graphs.

(2)   Since most of the readers and users expect a level twice as high to be twice as large, these types of graphs may be misleading to them because they do not understand what type of comparison is shown in graphs.

(3)   These types of graphs look very technical and discourage the readers and users to try to recognize and explain these graphs.

Because of the aforementioned problems and difficulties, though logarithmic line graphs provide a more complete depiction and explanation of the relationship between series of data, these graphs are not widely recommended and used to represent geographical data. These graphs are not suitable to display the geographical data in those fields and reports where common people are the targeted audience. These graphs are excellent for specialized purposes and are for audiences having adequate technical knowledge. In comparison, arithmetic line graphs are commonly and frequently used for the representation of geographical data because these graphs are simple and easy to understand.

Difference Between Arithmetic (Linear) and Logarithmic Line Graphs

The major differences between *arithmetic and logarithmic line graphs* are:

Arithmetic line graph	Logarithmic line graph
1. Arithmetic or linear scale is used on both the axes, i.e. on the X-axis (horizontal) and Y-axis (vertical) to represent the data	1. Logarithmic scale is used either on the X-axis (horizontal) or Y-axis (vertical) or both axes to represent the data
2. Arithmetic line graph can be used to represent any type of geographical data	2. Logarithmic line graph is commonly used when the range of values of the data set is very large
3. On an arithmetic scale, equal distances represent equal amounts or values, i.e. the values of a data series plotted on an arithmetic scale increase or decrease at a constant rate	3. On a logarithmic scale, equal distances represent equal ratios, i.e. the values of a data series plotted on a logarithmic scale increase or decrease at a constant ratio
4. Zeroes or negative values can be easily represented on arithmetic line graphs	4. Representation of zeroes or negative values on logarithmic line graphs is not possible

(continued)

(continued)

Arithmetic line graph	Logarithmic line graph
5. Representation of data on an arithmetic scale would produce a curving line, descending at a declining (getting lower) angle for a diminishing series of values and ascending at a rising (getting higher) angle for a growing series of values	5. A set of data plotted using a logarithmic scale on the y-axis will resemble a straight line, slanting up or down based on the nature of the data values. When the values increase or decrease at a constant rate, it will appear as a straight line
6. Arithmetic line graphs are very easy to read and understand to the common people because basic mathematical principles are applied	6. These types of graphs look very technical and create difficulties to understand to the common people. They are excellent only for the specialized audiences having adequate technical knowledge
7. These graphs only show the absolute changes of values but not the relative changes. Thus, these are not useful for comparing the relative changes (percentage) of values of geographical variables	7. Logarithmic line graphs provide a more complete depiction and explanation of the relationship between sets of data. These graphs are useful to identify and understand the relative changes (percentage) of values

Composite or Compound Line Graph

Sometimes the line graph shows the relationship between two or more than two variables or elements or facts called *composite or compound graph*.

Poly Graph

Poly graph is a multiple line graph in which two or more sets of variables are represented by distinctive lines (Fig. 2.11). It is frequently used for immediate comparison between several sets of variables, for instance, the death rates and birth rates of different states in a country; male and female literacy rate in different census years in a country (Table 2.5); proportion (%) of child, adult and old population in different census years in a country; amount of production of different crops (rice, wheat, maize, pulses etc.) in different years in a region etc. Generally, different variables are represented by different line patterns like a straight line (___), dotted line (......), broken line (- - -) or line of various colours (Fig. 2.11).

Band Graph

A *band graph* is practically a standard and aggregate line graph which shows the trends of values in percentage or numbers or quantity for successive time periods in both the total and its component parts (Table 2.6) by a series of lines drawn on the same frame (Fig. 2.12). *Band graph* shows how and in what proportion the component items constituting the aggregate are distributed. Different component items are represented one above the other and the intervening gaps between the

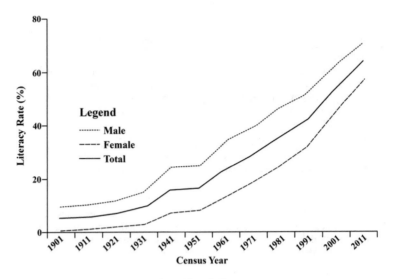

Fig. 2.11   Poly graph showing total, male and female literacy rates

Table 2.5   Worksheet for poly graph (Total, male and female literacy rates in different census years in India)

Census year	Literacy rate (%)			Scale selected	Literacy rate according to scale (cm)		
	Total	Male	Female		Total	Male	Female
1901	5.35	9.83	0.60	1 cm to 10% literacy rate	0.53	0.98	0.06
1911	5.92	10.56	1.05		0.59	1.06	0.1
1921	7.16	12.21	1.81		0.72	1.22	0.18
1931	9.5	15.59	2.93		0.95	1.56	0.29
1941	16.1	24.9	7.30		1.6	2.5	0.73
1951	16.67	24.95	7.93		1.7	2.5	0.79
1961	24.02	34.44	12.95		2.4	3.4	1.29
1971	29.45	39.45	18.69		2.9	3.9	1.87
1981	36.23	46.89	24.82		3.6	4.7	2.48
1991	42.84	52.74	32.17		4.3	5.3	3.2
2001	54.51	63.23	45.15		5.4	6.3	4.51
2011	64.32	71.22	56.99		6.4	7.1	5.7

Source Census of India

successive lines are filled by different colours or shades so that the graph looks like a series of bands (Fig. 2.12). When the differences in values in component parts are small, then the band graph becomes impressive representing the trends of their distribution but when the variations are too large then the band graph becomes less

Table 2.6 Worksheet for band graph (Production of different crops in India)

Year	Production (million tons)					Total (million tons)	Scale selected	Production according to scale (cm)				
	Rice	Wheat	Cereals	Pulses	Food grains			Rice	Wheat	Cereals	Pulses	Food grains
2004–05	83.1	68.6	185.2	13.1	198.4	548.4	1 cm to 100 million tons	0.83	0.69	1.85	0.13	1.98
2010–11	96.0	86.9	226.3	18.2	244.5	671.9		0.96	0.87	2.26	0.18	2.44
2011–12	105.3	94.9	242.2	17.1	259.3	718.8		1.05	0.95	2.42	0.17	2.59
2012–13	105.2	93.5	238.8	18.3	257.1	712.9		1.05	0.93	2.39	0.18	2.57
2013–14	106.7	95.9	245.8	19.3	265.0	732.7		1.07	0.96	2.46	0.19	2.65
2014–15 (4th Adv Est.)	104.8	88.9	235.5	17.2	252.7	699.1		1.05	0.89	2.35	0.17	2.52

Source Directorate of Economics and Statistics, Ministry of Agriculture and Farmers Welfare

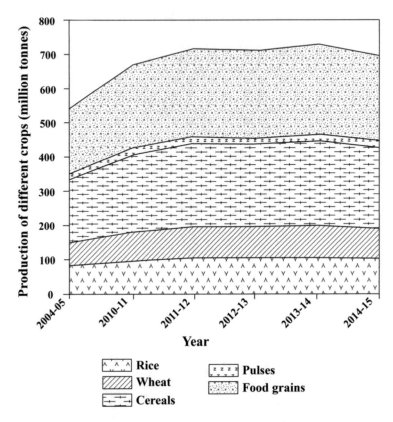

Fig. 2.12 Band graph showing the production of various crops in different years in India *Source* Directorate of Economics and Statistics, Ministry of Agriculture and Farmers Welfare

impressive and its legibility and clarity is marred. In the geographical study, band graph is useful for different purposes, including dividing the total crop production into different crops, total cost into component costs, total production by type of commodity or industries and other such relationships.

2.5.1.2   Closed Line Graph

Climograph

The concept and the idea of *climograph* was first conceived by J. Ball in the form of '*Climatological Diagrams*' (Singh and Singh 1991) and it was introduced by Griffith Taylor in the first half of the twentieth century (1949). The variations of world climatic conditions were summarized by Koppen using this graph. Again, J.B. Leighly explained the idea of Koppen to compare the climatic conditions of different parts of the world. Additionally, another two important types of climograph were

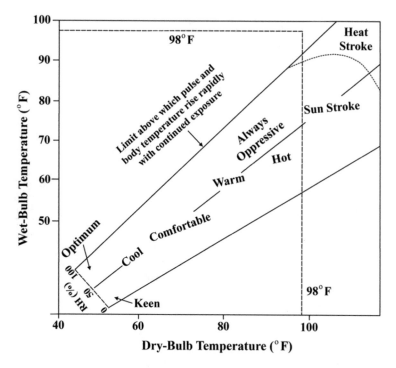

Fig. 2.13 USDA type of climograph

designed by the United States Department of Agriculture (U.S.D.A., 1941) and E.E. Foster (1944). The climograph was actually devised to show the scale of habitability for white settlers within the tropics.

Climograph of USDA Type (1941)

The United States Department of Agriculture devised a type of climograph in 1941 in which mean monthly wet-bulb temperatures (°F) is plotted against mean monthly dry-bulb temperatures (°F) on a referenced frame. Twelve points, each for one month, are obtained on the graph paper and the joining of these points results in a closed 12-sided polygon called climograph. Generally, this type of climograph is depicted to explain the climatic conditions with respect to human physiological comfort (Fig. 2.13).

Climograph of Foster Type (1944)

In 1944, E.E. Foster devised a type of climograph in which mean monthly temperatures (°F) is plotted against those of monthly precipitation (inches) on a referenced

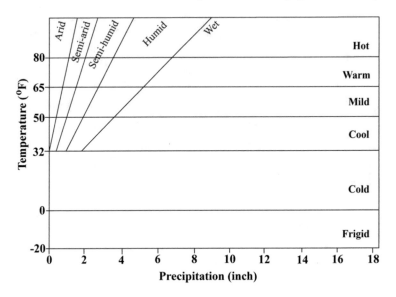

Fig. 2.14 The base frame of Foster's climograph

frame. Monthly precipitation (rainfall) is shown along the 'X'-axis (abscissa), graduated from 0 to 18 inch and the mean monthly temperature is shown along the 'Y'-axis
(ordinate), graduated from −20 to 100 °F (Fig. 2.14).

The reference frame has been divided into six temperature zones from bottom
to top, namely frigid zone (−20–0 °F), cold zone (0–32 °F), cool zone (32–50 °F),
mild zone (50–65 °F), warm zone (65–80 °F) and hot zone (more than 80 °F).
Additionally, the top four zones are divided into five sub-zones based on the amount
of precipitation, namely arid zone (0.32–1.03 inch), semi-arid zone (0.59–1.93 inch),
sub-humid zone (1.10–3.60 inch), humid zone (2.05–6.73 inch) and wet zone (more
than 2.05 inch and more than 6.73 inch) (Fig. 2.14). Each month is depicted by a
letter symbol and the joining of these points results in a closed 12-sided polygon
called climograph. This type of climograph is primarily used to depict the climatic
classification system proposed by C.W. Thornthwaite.

Climograph of G. Taylor (1949)

According to G. Taylor, *climograph* is a 12-sided polygon obtained from the graphical representation of wet-bulb temperature (°F) and the relative humidity (%) of
12 months of a particular place or station it corresponds to (Fig. 2.15 and Table 2.7).
The relative humidity is shown along the 'X'-axis (abscissa), graduated from 20 to
100% and the wet-bulb temperature is shown along the 'Y'-axis (ordinate), graduated from −10 to 90 °F. The 12 points (each for a month) are obtained on the graph
paper by plotting wet-bulb temperature against relative humidity for 12 months and

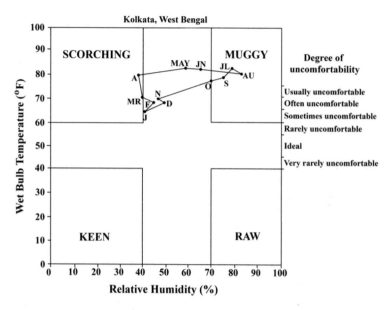

Fig. 2.15 Climograph showing the wet-bulb temperature and relative humidity of Kolkata (after G. Taylor)

Table 2.7 Monthly wet-bulb temperature (°F) and relative humidity (%) of Kolkata, West Bengal

Months	Jan	Feb	Mar	Apr	May	Jun	Jul	Aug	Sep	Oct	Nov	Dec
Wet-bulb temperature (°F)	64.7	68.5	70.5	79.8	82.9	82.4	80.8	82.6	80.4	78.1	68.9	68.5
Relative humidity (%)	41	45	40	39	59	67	83	80	76	70	46	49

joining of these points results in a 12-sided polygon called climograph. He marked four corners in the framework demonstrated by Keen (SW), Raw (SE), Muggy (NE) and Scorching (NW) (Fig. 2.15).

(a) **Keen**: Low wet-bulb temperature (below 40 °F) and low relative humidity (below 40%).

(b) **Raw**: Low wet-bulb temperature (below 40 °F) and high relative humidity (above 70%).

(c) **Muggy**: High wet-bulb temperature (above 60 °F) and high relative humidity (above 70%).

(d) **Scorching**: High wet-bulb temperature (above 60 °F) and low relative humidity (below 40%).

He also designed a tentative scale of discomfort and identified six categories regarding it:

(1) Very rarely uncomfortable: below 45 °F (40–45 °F)

Table 2.8 Mean monthly temperature and rainfall of Burdwan district, West Bengal

Months	Jan	Feb	Mar	Apr	May	Jun	Jul	Aug	Sep	Oct	Nov	Dec
Mean temperature (°C)	19.3	21.7	26.1	29.7	31.7	30.2	29.6	29.1	28.4	28.2	24.1	18.6
Rainfall (cm)	0.54	2.12	8.54	8.93	9.26	24.1	23.8	36.53	23.7	6.74	2.52	0

(2) Ideal: 45–55 °F
(3) Rarely uncomfortable: 55–60 °F
(4) Sometimes uncomfortable: 60–65 °F
(5) Often uncomfortable: 65–70 °F
(6) Usually uncomfortable: above 70 °F (70–75 °F).

The graph shifting towards the corners indicates the discomfortable characteristics of the climate. 'Scorching' and 'Keen' zones represent the hot desert and cold climatic characteristics, respectively, whereas the 'Muggy' region indicates the tropical humid climate (Fig. 2.15). The shape of the climograph is useful to understand the climatic character of a place or station. It is very simple and easy to compare the unknown climates with reference to the shape of the typical climograph (Singh and Singh 1991).

- Spindle-shaped climograph indicates the dry continental climate.
- Diagonal climograph represents the Mediterranean type of climate.
- Diagonal elongated climograph indicates the monsoon type of climate.
- Full-bodied climograph represents the ideal British type of climate.

Hythergraph

Hythergraph, a special form of climograph, was devised by Taylor (1949) to show the relationship between mean monthly temperature and mean monthly rainfall. It is drawn in the same way as in the case of climograph. Mean monthly rainfall is represented along the 'X'-axis *(abscissa)* and mean monthly temperature is shown along the 'Y'-axis *(ordinate)*. The 12 points (each for a month) are obtained on the graph paper by plotting mean monthly temperature against mean monthly rainfall for 12 months and joining of these points results in a 12-sided polygon called hythergraph. Table 2.8 shows the mean monthly temperature and rainfall of Burdwan district, West Bengal and these data are graphically represented using a hythergraph (Fig. 2.16).

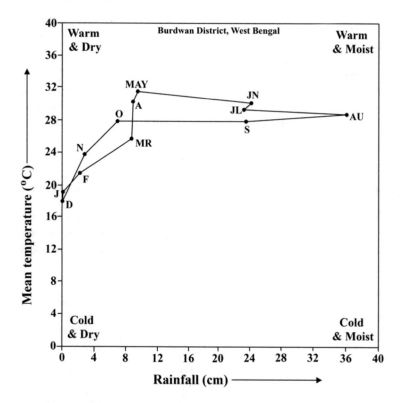

Fig. 2.16 Hythergraph showing the mean monthly temperature and rainfall of Burdwan district

Significance of Hythergraph

1. Hythergraph is principally used for comparing the climatic characteristics of different regions as affecting the cultivation of various crops, like rice, wheat, pulses, cotton etc.
2. It summarizes the basic climatic differences with respect to human activity, specifically in the context of settlement.

2.5.2 Tri-axial Graphs

Values of three geographical things or elements are represented along the sets of 'X', 'Y' and 'Z' axes on a reference frame. These graphs are very useful to represent three inter-related variables.

2.5.2.1 Ternary Graph

A *ternary graph* is an *equilateral triangular graph* that displays the proportion of three inter-related components or variables that sum to a constant (100%). Three components or variables are shown along three sides of the triangle, respectively. Each side of the triangle is graduated from 0 to 100% to represent tri-component data. The vertices of the triangle are given by (1, 0, 0), (0, 1, 0) and (0, 0, 1), i.e. each apex forms 0% on two scales and 100% on the third (Fig. 2.17).

Techniques and Principles of Representation of Data in Ternary Graph

Ternary graph becomes very useful whenever the data composed of three inter-related components or variables can be converted into percentages totalling 100. In this graph, data are represented by using *barycentric co-ordinates* (barycentric

Sub-triangle	Nature of distribution	Sub-triangle	Nature of distribution
I	Low B, Medium C & High A	IV	Low A, Medium C & High B
II	Low C, Medium B & High A	V	Low A, Medium B & High C
III	Low C, Medium A & High B	VI	Low B, Medium A & High C

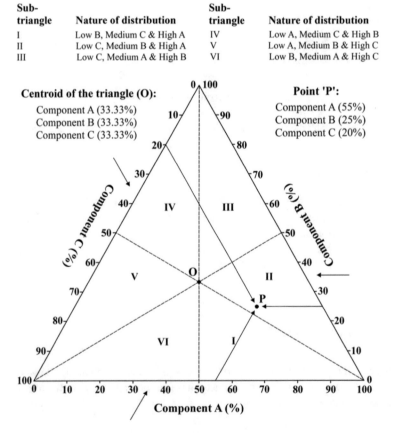

Fig. 2.17 Identification of position of points in ternary graph

Table 2.9 Database for ternary graph (Proportion of sand–silt-clay in sediments)

Stations	Proportion of sediment particles (%)		
	Sand	Silt	Clay
Kolaghat	68	25	7
Soyadighi	65	17	18
Anantapur	52	30	18
Pyratungi	60	12	28
Dhanipur	45	35	20
Geonkhali	54	16	30

co-ordinates are triples of numbers $[x_1, x_2, x_3]$) corresponding to amounts placed at the vertices of a reference triangle (say $\triangle ABC$). These amounts then determine the location of a point 'P', which is the geometric centroid of the three amounts and is identified with co-ordinates (x_1, x_2, x_3) (Fig. 2.17).

Since the values of three components all add up to 100%, all three values are plotted on the graph as a collection of points. The position of points within the triangular graph reflects the relative dominance of each component (Fig. 2.17).

Example

Data like age composition (young, adult and old), textural composition of soil or sediment (sand, silt and clay), occupational structure of population (primary, secondary and tertiary) etc. are suitable for representation in ternary graph.

Table 2.9 shows the proportion of sand, silt and clay in the sediment samples collected from the bed of Rupnarayan River at different sites (Kolaghat, Soyadighi, Anantapur, Pyratungi, Dhanipur and Geonkhali) and it is represented graphically in a ternary graph (Fig. 2.18). Types of sediment samples are easily understood by observing the location of points in the ternary graph.

A *ternary graph* was, however, found the most appropriate technique for the classification of a large number of Indian towns. Asok Mitra (1964) for the first time used the ternary graph for functional classification of towns in the 1961 census. His method of classification is based on the concept of dominant functions of a city. The seven census categories of workers were grouped into three broad non-agricultural categories, namely (1) industry, (2) trade and transport and (3) services. The percentages of three categories of towns are then plotted on a ternary graph and their position in the triangle was taken as the main determinant of their functional classification.

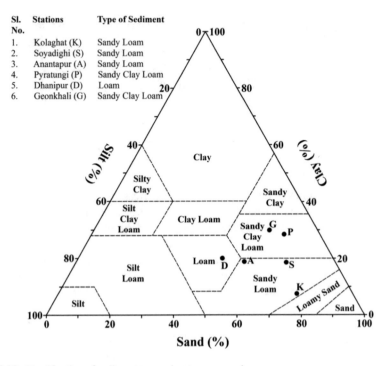

Sl. No.	Stations	Type of Sediment
1.	Kolaghat (K)	Sandy Loam
2.	Soyadighi (S)	Sandy Loam
3.	Anantapur (A)	Sandy Loam
4.	Pyratungi (P)	Sandy Clay Loam
5.	Dhanipur (D)	Loam
6.	Geonkhali (G)	Sandy Clay Loam

Fig. 2.18 Identification of sediment type using ternary graph

2.5.3 *Multi-axial Graphs*

The reference frame is composed of a network of evenly spaced lines radiating from the centre. The radiating lines are drawn at true azimuth on vector graph. Values are plotted along the radiating lines and the obtained points are then joined.

2.5.3.1 Radar or Spider or Star Graph

A *radar graph*, also called *spider graph* or *star graph*, is the graphical representation of the *multivariate data* (three or more variables) in the form of a two-dimensional graph consisting of a sequence of equi-angular spokes or axes starting from the same point called radii, each representing a different variable. The length of each spoke is proportional to the quantity of the variable for the data point with respect to the maximum quantity of the variable across all data points. The spokes are then joined with a line of a selected colour or pattern in the form of a shaded polygon to represent each category, creating a star-like shape with points equal to the number of categories (Fig. 2.19).

A *radar graph* provides the user with numerous visual comparisons by portraying multivariate data with different variables. In other words, if we want to understand

Fig. 2.19 Radar graph
(Production of different
crops)

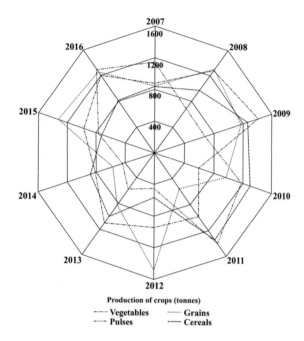

Production of crops (tonnes)
---- Vegetables   ······ Grains
---- Pulses   ---- Cereals

how multiple data points interact with each other and make a comparison against
another set of multiple data points, then a radar graph is one of the best ways.
This graph is mainly drawn to display the continuous diurnal, monthly or annual
rhythm of different geographic variables. For example, hourly data of atmospheric
humidity, atmospheric temperature, sunshine and soil temperature; monthly data of
atmospheric humidity, rainfall, atmospheric temperature; and yearly data of production of different crops, industrial goods etc. can be easily represented in a radar
graph.

Methods of Construction

1.   A suitable scale is at first selected to represent the data, and the required number
     of concentric circles is drawn at regular intervals (Fig. 2.19).
2.   The required number of equi-angular radial straight lines (axes) of corre-
     sponding lengths is drawn from the centre (i.e. the point of origin). Each axis
     shares the same divisions and scale, but the method in the range of variable
     values maps to this scale may be different between the represented variables.
3.   The values from a single observation are represented along each axis and joined
     to form a polygon to make the graph more easily readable and understandable
     (Fig. 2.19).
4.   More number of observations can also be placed in a single graph with the help
     of multiple polygons.

Table 2.10 Data for radar graph (Production of different crops in different years)

Year	Production of different crops (tonne)			
	Grains	Pulses	Vegetables	Cereals
2007	850	1143	885	847
2008	1284	691	1295	980
2009	1170	680	1409	1272
2010	974	1182	599	1310
2011	554	1405	972	1354
2012	1491	453	937	568
2013	820	572	1085	1004
2014	584	439	802	885
2015	1319	762	974	726
2016	1270	1277	1214	815

5.   Overlay these polygons and adjust the opacity of every polygon for each of the observations. Hence, an hourly graph is represented by a 24-sided polygon and a monthly graph by a 12-sided polygon.

Steps of Drawing Radar Graph in Microsoft Excel

Step 1: Insert the data in a suitable format.
Step 2: Go to Insert tab → Other Charts → Select Radar with Marker Chart. A blank radar graph will be inserted.
Step 3: Right click on the blank graph and click on select data.
Step 4: Click on Add button
Step 5: Select Series name as Grains (for the following example) and Series value as production values (Table 2.10) and click Ok.
Step 6: Repeat the same procedure for all the remaining data. After this, click on Ok and a graph will be inserted (Fig. 2.19).
Step 7: Format the graph according to your need.

How to Understand the Radar Graph

Like the column or bar graph in the spider graph we also have 'X' and 'Y' axes. The X-axis is nothing but the extreme end of the radial line (spider) and each step of the radial line is considered as Y-axis. Zero (0) point of the radar graph starts from the centre of the wheel. Towards the margin of the spike, a point arrives, and the higher is the value.

Interpretation of the Graph

- By having a look at the spider graph (Fig. 2.19) we can understand that in 2012 the production of grains was the highest among all the 10 years productions. In 2011, the production of grains was the lowest.
- In the case of pulses, the highest production was in 2011 and the lowest production was in the year 2014.
- In the case of vegetables, the highest production was in the year 2009 and the lowest production was in 2010.
- In the case of pulses, the highest production was in the year 2011 and the lowest production was in the year 2012.

Advantages of Using Radar Graph

Radar or spider graphs are frequently used in the geographical analysis for the comparison of the distributions along the radial lines of different directions of frequency and index data to compare two or more areas. The major advantages of using this graph are as follows:

1. The radar graph is more suitable when the absolute values aren't critical for a user but the whole graph tells some story.
2. Radar graphs are very useful for strikingly showing outliers and commonality, or when one graph is larger in every variable than another.
3. Several attributes become easily comparable, each along their own axis, and their overall variations are clear by the shape and size of the drawn polygons.
4. In radar graph, many variables can be easily shown next to each other whilst still giving each variable the identical resolution.
5. Radar graphs are more effective when there is the need to compare the performance of one thing to a standard or a group's performance. For example, if one has a radar graph portraying data about the average quantity of production of crops in different regions, one could easily superimpose another polygon representing a particular crop production data in order to quickly observe how that crop compares to average crop production in each region (Fig. 2.19).

Limitations

The major disadvantages of this graph are:

(1) The comparison of data on a radar graph becomes difficult once there are more than two webs on the graph.
(2) When too many variables are represented along different axes, it creates the crowding of data.
(3) Though there are several axes which have gridlines joining them for indication, problems arise when observers seek to compare the values along different axes.

(4) Each axis of a radar graph shares a common scale, which means that the range of values of each variable requires to be mapped based on this shared scale in a different way. The way of mapping these variables is not understandable in most cases, and can even be misleading.

(5) Another important problem is that observers could potentially think that the area of the polygons is a very important thing to consider. But, the shape and area of the polygons can vary largely based on how the axes are placed around the circle.

Alternatives to radar graphs are bar graphs and parallel coordinate graphs.

2.5.3.2 Polar or Rose Graphs

A *polar or rose graph* is the graphical representation of the direction as well as magnitude or quantity of different phenomena or variables, especially geographical in nature. Phenomena or variables characterized by direction and distance from a specific point of origin can be plotted on polar graphs (Figs. 2.20 and 2.21). They are analogous to radar graphs, but as a substitute for any variable. They exclusively emphasize geographical phenomena.

Fig. 2.20 Wind rose graph showing the percentage of days wind blowing from different directions

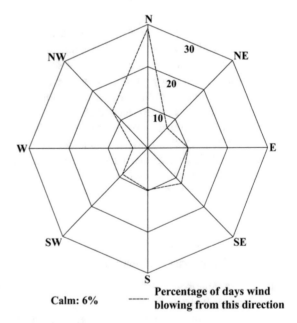

Calm: 6%   - - - - - **Percentage of days wind blowing from this direction**

Fig. 2.21 Polar graph showing the number of corries facing towards different directions

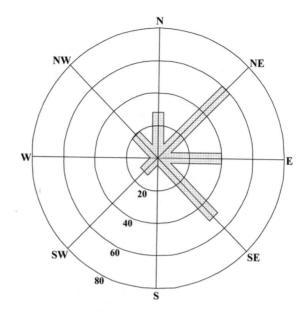

Principles and Methods of Construction

In polar or rose graph, values are plotted as radii from a point of origin (pole) with the help of a polar coordinate system. This graph draws the '*X*' and '*Y*' co-ordinates in each series as (*theta [θ]*, *r*) (discussed in types of co-ordinate section), where theta is the amount of rotation from the origin (vector angle) and '*r*' is the distance from the origin (radius vector). The outer values in the circle always represent the degrees in the circle. Data '*X*' holds the *x*-axis position in degrees and data '*Y*' holds the position of each phenomenon of the variable on the *y*-axis (Figs. 2.20 and 2.21). These graphs are especially useful where vector values are involved.

Polar or rose graphs are useful for the analysis of various geographical data containing magnitude and direction values. These graphs are generally used to represent the direction, magnitude and frequency of ocean or wind waves, the direction of facing of cirques or corries, the orientation of the long axes of pebbles or boulders etc.

Wind rose graph is the most frequently used polar graph in geographical analysis. Meteorologists, climatologists and geographers use wind rose to graphically display wind speed and wind direction at a particular location over a defined observation period (Table 2.11 and Fig. 2.20). Wind rose can be prepared for month-wise, season-wise or yearly as required. It typically uses 16 cardinal directions, such as north (N), north-east (NE), south (S), east (E) etc., although they may be sub-divided into as many as 32 directions. In terms of the measurement of angle in degrees, north corresponds to 0°/360°, east to 90°, south to 180° and west to 270°.

Figure 2.20 shows the percentage of days wind blowing from different directions. It is clear from the graph that only 6% of winds are in a calm condition. About 29% of

Table 2.11 Percentage of days wind blowing from different directions

Wind direction	Percentage of days wind blowing from this direction
North	29
North-east	7
East	10
South-east	12
South	10
South-west	9
West	4
North-west	13
Calm	6

Table 2.12 Data for polar graph (The orientation of corries in a glacial region)

Orientation of corries	Number of corries facing towards this direction
North	30
North-east	60
East	40
South-east	50
South	0
South-west	10
West	0
North-west	20

winds are blowing from the northerly direction, followed by 13% from north-west, 12% from south-east, 10% from south, 9% from south-west etc. So, there is enough variability in the directions of wind blowing.

Table 2.12 and Fig. 2.21 show that most of the corries in the glacial region are facing towards north, east and south-east directions. It is evident from the figure that the faces of 60 corries are in the north-easterly direction, 40 corries are in the easterly direction and 50 are in the south-easterly direction.

Advantages and Disadvantages of the Use of Polar or Rose Graph

Advantages

1. Multiple sets of data can be easily compared.
2. Lots of data can be represented on a single graph.
3. Easy to understand and interpret.
4. Individual components within the graph can be easily compared.

Disadvantages

1. Linking the data and statistical tests is difficult.
2. It is hard to spot anomalies.
3. Difficult to consider a suitable scale.

2.5.4 Special Graphs

2.5.4.1 Scatter Graph

Scatter graph is the simplest and easiest way to show the relationship between two variables (*bi-variate data*) at a glance. Bi-variate data is the data that deals with the simultaneous measurement of two variables that can change and are compared to find the relationships. In bi-variate data, one variable is influenced by another variable and thus bi-variate data has an independent (X) and a dependent variable (Y) (Table 2.13). It is because of the fact that the change of one variable depends on the change of the other. An independent variable is a part of data or condition in an experiment that can be changed or controlled. A dependent variable is a part of data or condition in an experiment that is affected or influenced by an external factor, most frequently the independent variable. Bi-variate data can be easily represented in a scatter graph to understand the type and nature of co-relation that exists between them. In this method, the independent variable is shown along the 'X'-axis and the dependent variable is shown along the 'Y'-axis. Scattered points are obtained by putting the values of Y with respect to X. In case of a trend or correlation, a 'line of best fit' can then be drawn within a scatter graph (Fig. 2.22).

For example, the relationship between height from mean sea level and the number of settlements, the relationship between basin area and run-off etc. can be easily represented using scatter graph.

Table 2.13 Database for scatter graph (The distributions of air temperature in the month of April around an urban area)

Distance from CBD (km) [X]	Air temperature (°C) [Y]
1.3	41.25
3.7	41.02
5.2	40.39
7.1	39.87
9.7	39.58
11.5	39.01
17.5	38.09
18.2	37.73
21.7	35.25
25.7	32.80

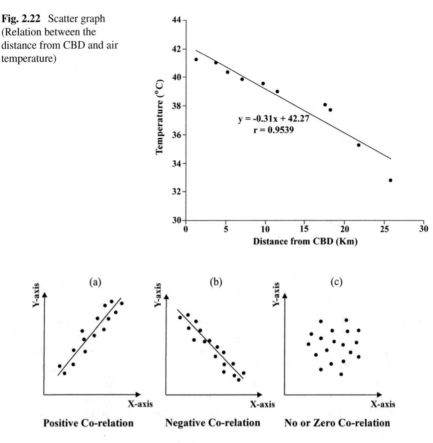

Fig. 2.22 Scatter graph (Relation between the distance from CBD and air temperature)

Fig. 2.23 Positive, negative and no co-relation

The location and orientation of points in the scatter graph indicate the type and nature of co-relation between variables.

Positive, Negative and Zero Co-relation

When the points are oriented from lower left to upper right then it indicates the *positive co-relation* between the *variables* (Fig. 2.23a). Here, two sets of data or variables steadily tend to move together in the same direction, i.e. an increase in the value of one set of variables causes an increase in the value of another set of variables and vice versa. For example, distances travelled and amount of transport cost, the slope of land and rate of soil erosion, income and expenditure of a family etc. are positively co-related. When the values of both the variables tend to move together in the same direction with a constant proportion then it is known as a perfect positive correlation. In this co-relation, all the plotted points lie on a straight line that rises

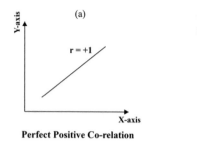
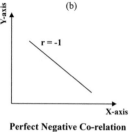

Fig. 2.24 Perfect positive and negative co-relation

from the lower-left corner to the upper-right corner. Numerically, it is indicated as +1 ($r = +1$) (Fig. 2.24a).

On the other hand, when two sets of data or variables steadily tend to move together in the opposite direction, i.e. an increase in the value of one set of variables causes a decrease in the value of another set of variables and vice versa, then it is called *inverse or negative co-relation* (Fig. 2.23b). For example, height from mean sea level and the number of settlements, distance from forest area and amount of organic matter in the soil, price of a commodity and its demand etc. are negatively co-related. When the values of both the variables tend to move together in the opposite direction with a constant proportion, then it is known as perfect negative co-relation. In this co-relation, all the plotted points lie on a straight line falling from the upper-left corner to the lower-right corner. Numerically, it is indicated as -1 ($r = -1$) (Fig. 2.24b).

When one set of variables does not change even with the change of another set of variables (change in one variable does not depends on the change of another variable), then the relationship between them is called *zero co-relation* or *non-co-relation,* i.e. no co-relation exists between variables (Fig. 2.23c). For example, marks in physics and marks in geography, the height of persons and their intelligence etc.

Linear and Nonlinear or Curvilinear Co-Relation

Linear or nonlinear co-relation is a function of the constancy of change of ratio between two variables. In *linear co-relation,* the amount of change in one variable maintains a constant ratio to the amount of change in the other variable, i.e. the ratio of change of values between two variables is equal. The points obtained from the plotting of the values of one variable with respect to the other on a graph will move around a line (Fig. 2.25a).

In *nonlinear or curvilinear co-relation,* the amount of change of variables is not constant, i.e. the ratio of change of values between two variables is unequal. The points obtained from the plotting of the values of one variable with respect to the other on a graph will move around a curve (Fig. 2.25b).

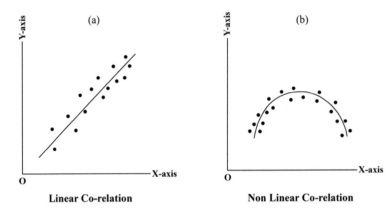

Fig. 2.25 Linear and nonlinear co-relation

2.5.4.2 Ergograph

The term *'ergograph'* was first used by Dr. Arthur Geddes of the University of Edinburgh. An ergograph is a special kind of multivariate graph which represents the relationship between season, climatic elements and cropping patterns (human activities). Various stages of the cycle of plant growth, i.e. sowing, growing, flowering, maturing, harvesting etc., intimately corresponds to seasonal characteristics of weather conditions. Variation of seasons are manifested by different climatic characters and cropping patterns. The time of maturity of different crops varies from one another. Some crops are annual, some are bi-annual and some may require only a few months to be matured. Ergograph can be drawn either by the Cartesian co-ordinate method (rectangular form) or by the polar co-ordinate method (circular form).

In the Cartesian co-ordinate method or rectangular form, different climatic elements like mean monthly temperature, rainfall, relative humidity etc. (Table 2.14) are marked along the 'Y'-axis (vertical axis) in the form of polygraphs. However, the monthly rainfall is generally represented by vertical bars. The 12 months are plotted along the 'X'-axis (horizontal axis). Below the horizontal axis (primary baseline), a crop calendar is drawn in the form of rectangles on a selected scale to represent the acreage of different crops (Fig. 2.26 and Table 2.15). The length of each rectangle must correspond to the growing season of the crop while

Table 2.14 Data for ergograph (Monthly temperature, relative humidity and rainfall of Howrah, West Bengal)

Months	Jan	Feb	Mar	Apr	May	Jun	Jul	Aug	Sep	Oct	Nov	Dec
Temperature (°C)	16	18	30	32	33	30	28	28	27	25	19	17
Relative humidity (%)	51	48	54	60	65	78	80	81	75	70	57	49
Rainfall (cm)	1.5	5.3	7.8	6.9	12.4	22	23	34.3	24	7.5	3.9	0.8

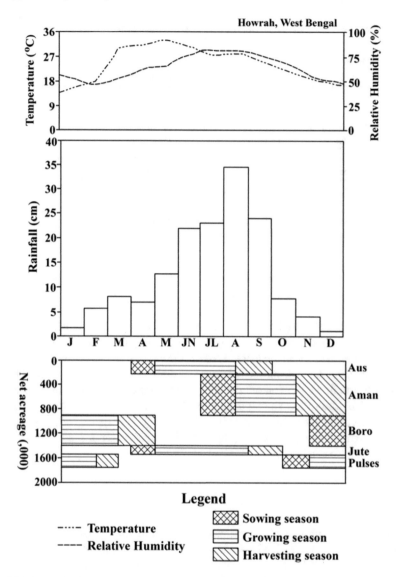

Fig. 2.26 Ergograph showing the relation between seasons, climatic elements and cropping patterns of Howrah, West Bengal

the breadth of them will be directly proportional to the crop acreage based on the selected scale. Again, each rectangle may be divided into different parts to indicate the time periods of various stages of crops grown (Fig. 2.26).

Table 2.15 Data for ergograph (Net acreage of different crops and their growing seasons of Howrah, West Bengal)

Crops	Seasons			Net acreage (,000)
	Sowing	Growing	Harvesting	
Aus	April	May to mid of August	Mid of August to September	200
Aman	July to mid of August	Mid of August to October	November to December	650
Boro	Mid of November to December	January to mid of March	Mid of March to mid of April	500
Jute	April	May to August	September to mid of October	120
Pulses	Mid of October to mid of November	Mid of November to mid of February	Mid of February to mid of March	220

Polar Co-ordinate or Circular Ergograph of A. Geddes and G.G. Ogilvie (1938)

A. Geddes and G.G. Ogilvie (1938) developed *polar co-ordinate* or *circular form of ergograph* to show the continuous rhythm of seasonal activities (Table 2.16) in which 12 months of a year are marked around the circumference of the circle, forming 30° sectors (Fig. 2.27). Concentric curves are drawn to show the nature of activities done each month and the amount of time (hours per day) assigned to each type of activity. The time scale, ranging from 0 to 24 h per day, is a square root scale and is represented along the radius of the circle (Fig. 2.27). This type of ergograph is also popularly known as a polar strata graph or polar layer graph or polar line graph (as the data form 'bands' on the graph).

2.5.4.3   Ombrothermic Graph

Climatic graphs summarize the trends in temperature and precipitation for no less than 30 years. They are likely to establish the relationship between temperature and precipitation and determining the span of dry, wet and extremely wet periods. *Ombrothermic graph,* also called Walter Lieth graph, is an important climatic graph used to compare the average dryness and wetness of a place. The data of *ombrothermic graph* must be the average for no less than 30 years. This graph was first designed and used by French bio-geographer and naturalist, Marcel-Henri Gaussen to graphically depict the mean monthly temperature and monthly precipitation of a place.

Table 2.16 Database for circular ergograph (Rhythmic seasonal activities)

Time devoted to various activities (hour/day)	Months											
	Jan	Feb	Mar	Apr	May	Jun	Jul	Aug	Sep	Oct	Nov	Dec
Domestic works	4.5	5	5	5.5	5	5.5	6	5.5	5.5	5.5	5	4.5
Agricultural activities	5	4.5	6	5.5	6.5	6	6	6.5	5.5	6	6	6
Animal husbandry	1.5	1	2	2	1.5	0.5	0.5	0.5	1.5	1.5	1	1.5
Fishing	0.5	0.5	1	1.5	1.5	2.5	3	2	1.5	1	1	0.5
Entertainment	1.5	2	1	1	1.5	1.5	1	1.5	2	2.5	2	1.5
Others including sleep	11	11	9	8.5	8	8	7.5	8	8	7.5	9	10

Table 2.17 Data for ombrothermic graph (Average temperature and rainfall of Purulia district, West Bengal)

Months	Jan	Feb	Mar	Apr	May	June	July	Aug	Sep	Oct	Nov	Dec
Average rainfalls (mm)	13	24	21	26	45	223	281	300	259	82	10	4
Average temperatures (°C)	18.8	21.9	26.8	31.6	33.3	31.2	28.3	28.1	28	26.5	22.2	19.4

Fig. 2.27 Circular ergograph showing the rhythm of seasonal activities (after A. Geddes and G.G. Ogilvie 1938)

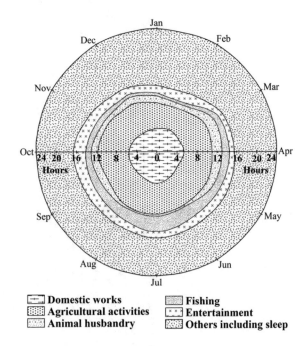

Domestic works   Fishing
Agricultural activities   Entertainment
Animal husbandry   Others including sleep

Principles and Methods of Construction

(1) The mean monthly temperature (°C) and monthly rainfall (mm) of Purulia district, West Bengal (Table 2.17) have been represented by Ombrothermic graph (Fig. 2.28). For the drawing of this graph, months of the year are shown along the x-axis while one y-axis (y_1) represents the mean monthly temperature and another y-axis (y_2) represents the total monthly precipitation.

(2) The x-axis of the graph should begin with the coldest month of the year. In the case of the places located in the Northern Hemisphere, the x-axis should start with January, whereas in the Southern Hemisphere, it should start with the month of July.

(3) Mean monthly temperature and monthly precipitation should be expressed in degree centigrade (°C) and millimetres (mm), respectively. The selection of the scales (y_1 and y_2) is very important and it should follow the following relationship:

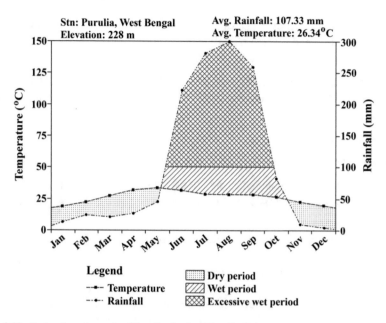

Fig. 2.28 Ombrothermic graph of Purulia district, West Bengal

$$\text{Temperature (T)} = \frac{1}{2} \times \text{Precipitation (P)}$$

For example, the mean monthly temperature of 5 °C on the y_1-axis have to be equal to 10 mm of total monthly precipitation on the y_2-axis.

(4) The selection of scales on both the axes of this graph is based on the Gaussen-Bagnouls Aridity Index (Sarkar 2015):

   (i) Precipitation more than three times the temperature (P > 3 T) indicates the wet period.
   (ii) Precipitation between two times and three times of the temperature (3 T > P > 2 T) indicates the semi-wet period.
   (iii) Precipitation less than two times the temperature (P < 2 T) indicates the arid or dry period.

(5) Generally, the precipitation and temperature curves are represented in blue and red lines, respectively.

(6) When the precipitation curve lies below the temperature curve then it indicates a period of dry condition or xeric period, but if the precipitation curve lies above the temperature curve then it indicates a period of wet condition. When the precipitation curve exceeds 100 mm then it signifies a period of an excessive wet condition (Fig. 2.28).

(7)   The station name and its elevation should be mentioned in the top left, average temperature and average rainfall in the top right, and extremes of temperature in the second line should be shown.

Inference: The station has a dry span between November and May and a wet span from June to October. Between the month of June and September, it is an excessive wet period.

Demerits of this graph

The major demerits of this graph are as follows:

(1)   It lays on a scale that applies only to the mid-latitude climates.
(2)   It can be only constructed and read by using the metric system.

2.5.4.4   Water Balance or Water Budget Curve

In nature, water is almost in a constant motion due to the changes of its state from liquid to solid or vapour in appropriate environments. Law of mass conservation signifies that within a particular area for a specific period of time, the inflows and outflows of water are equal, including any change of storage of water in the concerned area, i.e. the water coming into an area has to depart the area or be stored within the area. The supply of groundwater in an area indicates whether a stage is one of water utilization, deficiency, recharge or surplus.

Water balance techniques are very important for the solution of various theoretical and practical hydrological problems and disasters. This approach helps us to evaluate the water resources quantitatively and their transformation due to the influence of human activities. Detailed knowledge about the water balance structure of river basins and groundwater basins offers a platform to make various hydrological projects valid for the control, redistribution and rational use of water resources with respect to time and space (Sokolov and Chapman 1974).

Formulation of Water Balance Techniques

Techniques of water balance estimation can be formulated by the following parameters and equations:

(A)   **Gains**: Precipitation (P)
(B)   Soil moisture recharge or storage (R)
(C)   **Losses**: Utilization (U) and evapotranspiration

    (a)   Actual evapotranspiration (AE)
    (b)   Potential evapotranspiration (PE)

Simple water balance

1.   Environments with abundant moisture condition

Table 2.18 Water need and supply (mm) of a region (field capacity: 100 mm)

Month	Jan	Feb	Mar	Apr	May	June	Jul	Aug	Sep	Oct	Nov	Dec
Supply of water (mm)	125	105	117	128	125	96	82	81	80	72	102	103
Need of water (mm)	6	12	33	66	108	149	170	156	113	63	20	12
Supply minus need	+119	+93	+84	+62	+17	-53	-88	-75	-33	+9	+82	+91
Water budget section	S	S	S	S	S	U	U/D	D	D	R	R	R/S

$$P > PE, \text{ thus } AE = PE$$

2.   Environments with inadequate moisture condition

$$P < PE, \text{ thus } AE < PE$$

3.   Environments with seasonal moisture condition.

In seasonal moisture environments, calculation of the monthly water budget is done based on the following parameters:

a.   Precipitation (P)
b.   Potential evapotranspiration (PE)
c.   Actual evapotranspiration (AE)
d.   Change in water storage (ΔST)
e.   Difference between P and PE (P − PE)
f.   Deficiency of water (D)
g.   Soil moisture storage up to field capacity (ST)
h.   Surplus of water (S): After attaining field capacity (ST), excess precipitation (P) is available as surplus.

A water balance equation is simply expressed as follows:

$$P = Q + E \pm \Delta ST \tag{2.15}$$

where P is precipitation, Q is run-off, E is evaporation and ΔST is the surface, sub-surface and groundwater storage.

In seasonal moisture environments, in the annual cycle of water balance estimation, a period of soil moisture recharge (R) is followed by a period of water surplus (S), and subsequently, a period of soil moisture utilization (U) is followed by a period of water deficiency (D) (Sutcliffe et al. 1981) (Table 2.18). The surplus of water comprises both surface run-off and groundwater storage.

Procedures for Determining the Status of Water Availability

In an annual cycle of water balance estimation, the periods of *soil moisture recharge (R), water surplus (S), soil moisture utilization (U) and water deficiency (D)* are identified by the following procedures:

(a)   In Table 2.18, month-wise water need and supply of a region for a theoretical one-year period is shown, assuming the field capacity of the soil being 100 mm. In all the months (January–May), the values of 'Supply minus need' are positive (supply of water is more than need) but from the month of June to September, the negative values (need of water is more than supply) indicate that more water will be withdrawn from underground than that will be recharged. Therefore, June is the month from which the utilization of storage water starts. In the 'Water budget section' of Table 2.18, write 'U' for the utilization of water in June.

(b)   Utilization of water continues until the total of the negative values becomes 100 (field capacity of the soil is 100 mm). If we add the values −53 and −88 for the months of June and July, then the total value exceeds 100. It indicates that July is the month of transition from water utilization to deficiency. Therefore, in the 'Water budget section' write 'U/D' for the month of July (Table 2.18).

(c)   In the 'Water budget section' write 'D' for all the months (in Table 2.18, for the months of August and September) to indicate deficiency of water (the need for water is more than supply) until the value becomes positive again. The positive values (supply of water is more than the need) indicate the recharge of water into the ground and it will be marked by the letter 'R' in the 'Water budget section' (October and November in Table 2.18).

(d)   Again, the recharge of water is converted to water surplus when the total of the successive positive values exceeds 100 (field capacity of the soil). In the month of December, the total of positive values exceeds 100 (9 + 82 + 91 = 182) indicating the time of transition from recharge to surplus of water. Therefore, write 'R/S' in the 'Water budget section' for that month. Then all the subsequent positive value is considered as the surplus of water and the letter 'S' is written in the 'Water budget section' (months of January–May in Table 2.18).

In this way, we will be able to easily determine the status of water availability of any region for all the months in a year and rational decisions can be made on how much and when water should be allocated for different purposes.

The surplus and deficiency of water of the sample study area during a normal rainfall year are shown at monthly intervals in Table 2.19 and Fig. 2.29. Month-wise rainfall (precipitation, P) and potential evapotranspiration (PE) have actually been superimposed in Fig. 2.29. This figure reveals that from the month of January to April amount of precipitation (P) exceeds potential evapotranspiration (PE), which is an indicative of surplus of water during this period. Then the potential evapotranspiration exceeds precipitation from the month of May to September. Consequently, the use

Table 2.19 Water budget estimation for a sample study area (elevation: 12 m; field capacity: 102 mm)

Month	Jan	Feb	Mar	Apr	May	June	Jul	Aug	Sep	Oct	Nov	Dec	Total
P (mm)	148	123	103	67	55	50	35	35	61	119	152	170	1118
PE (mm)	13	21	29	39	58	76	89	82	70	50	24	14	565
P − PE	+135	+102	+74	+28	−3	−26	−54	−47	−9	+69	+128	+156	
ΔST	0	0	0	0	−3	−26	−54	−19	0	+69	+33	0	
ST (mm)	102	102	102	102	99	73	19	0	0	69	102	102	
AE (mm)	13	21	29	39	58	76	89	54	61	50	24	14	528
D (mm)	0	0	0	0	0	0	0	28	9	0	0	0	37
S (mm)	135	102	74	28	0	0	0	0	0	0	95	156	590

of stored soil water was started and complete utilization of the stored water causes a deficit of water from the middle of August to September. The perpendicular line (R) drawn on the graph (Fig. 2.29) represents the complete utilization of stored soil moisture. Again, precipitation exceeds potential evapotranspiration from the month of October and continues up to April. A part of this water surplus compensates for the loss of soil moisture or it triggers the recharge of water which is completed by the middle of November. The perpendicular line (U) drawn on the graph (Fig. 2.29) represents that the field capacity (102 mm) has been reached, i.e. the soil moisture has been fully restored. The excess water obtained after saturation level of the field capacity is attained and termed surplus of water (from the middle of November to April) (Fig. 2.29).

Applicability of Water Balance Estimation

1. Water balance estimation is a useful method to assess the present status and trends of the availability of water resources in a region for a particular period of time.
2. Estimation of water balance assesses and improves the validity of visions, scenarios and various strategies which strengthens the procedures of decision-making for the proper management of water.

2.5.4.5 Hydrograph

A *streamflow or discharge hydrograph* at any point on a stream is a graph showing the flow rate as a function of time at that point. In this graph, the discharge is plotted on the y-axis (ordinate) and time is on the x-axis (abscissa) (Fig. 2.30). The units of time may be in minutes, hours or days, and the rate of flow (discharge) is generally expressed in cubic meters per second (cumec) or cubic feet per second (cusec). Thus, the hydrograph is an important graphical representation of the topographic

Water Balance Graph of a Sample Study Area
(Elevation-12m & Field capacity-102mm)

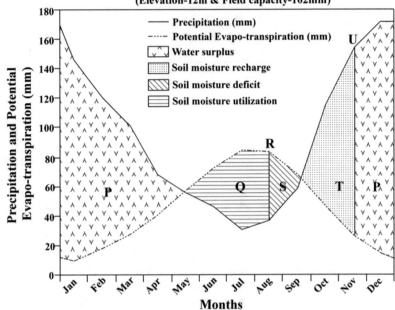

Months

P: Precipitation is greater than potential evapo-transpiration. Full storage of soil water leads to water surplus for the use of plants, run-off and ground water recharge.

Q: Potential evapo-transpiration is greater than precipitation. Storage water is being used up by plants or lost by evaporation (Soil moisture utilization).

R: The soil moisture storage is now used up. Any precipitation will be absorbed in soil rather than produce run-off. River levels will fall or may dry upcompletely.

S: Water deficiency due to utilization of storage water and potential evapo-transpiration is greater than precipitation. Plants must adupt to survive and crops must be irrigated.

T: Precipitation is greater than potential evapo-transpiration. Again, water will be recharged in the soil (Soil moisture storage).

U: Storage of soil water is full. Field capacity has been attained. Additional rainfall will percolate down to the water table and increase the ground water level.

Fig. 2.29 Water balance curve of a sample study area

and climatic characteristics which control the inter-relationship between rainfall and run-off of a particular drainage basin (Chow 1959). Though two types of hydrographs are particularly important: the annual hydrograph and the storm hydrograph, but it is more useful in hydrology to consider a hydrograph for a certain storm event (storm hydrograph).

A *storm hydrograph* reflects the influence of all physical characteristics of the river basin and, to some extent, also reflects the characteristics of the storm causing the hydrograph. A hydrograph can be considered a thumbprint of a drainage basin. The shape of a hydrograph mainly depends on the rate of transfer of water from

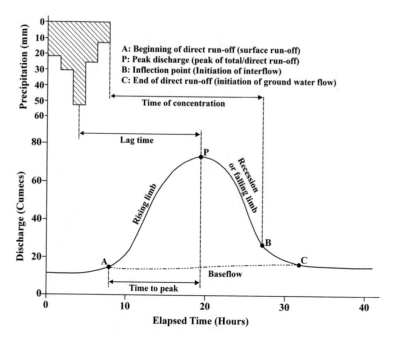

Fig. 2.30 Elements of a hydrograph

different parts of the river basin to the gauge station. No two river basins produce the same hydrographs for the same storm. Hydrographs from similar river basins may be similar, but not the same. In the same way, no two storms generate identical hydrographs from the same basin.

Elements of the Hydrograph

The elements of the hydrograph are:

(1) **Rising limb**: As surface run-off reaches the gauge station, the water begins to increase in the channel. With the progress of time, more and more surface run-off reaches the gauge and the water in the channel continues to rise until it reaches a maximum discharge, recorded as the maximum gauge height. The rising portion of the hydrograph indicated by the rising stage is called the rising limb (Fig. 2.30). The rising limb graphically represents increasing discharge over time as the limb rises and discharge increases.

(2) **Crest and peak discharge**: The time interval of the greatest discharge at the peak of the hydrograph is called the crest (Fig. 2.30). It may be of a short time duration represented as a sharp peak or of a fairly long time duration represented as a flat peak. The crest does not necessarily represent an equal volume of discharges, rather it represents a zone having nearly equal highest

discharges. The greatest discharge within the crest is called peak discharge (Fig. 2.30) and it is of primary interest in hydrologic design.

(3) **Recession or falling limb**: The segment of the hydrograph after the peak is called recession limb or falling limb (Fig. 2.30). It indicates the decrease of discharge as water is withdrawn from the river basin storage after rainfall ceases. The steepness of the recession limb represents the rate of draining of water from the basin area.

(4) **Point of inflection**: The point on the recession limb indicating the end of storm flow (i.e. quick flow or direct run-off) and the return to groundwater flow (i.e. *base flow*) is known as the **point of inflection** (Fig. 2.30). In other words, it is the point on the recession limb of the hydrograph where the steepness of the slope of the graph starts to decline. It indicates the point where the base flow becomes dominant to the total flow than the quick-response run-off.

(5) **Time to peak**: It is the time interval from the beginning of the rising limb (beginning of the increase of discharge at the gauge station) to the peak discharge (Fig. 2.30). It is largely controlled by the characteristics of the drainage basin like travel distances, drainage density, channel slope, channel roughness, soil infiltration capacity etc. The distributional pattern of rainfall over the basin area is very important to alter the time to peak in a hydrograph. For example, the hydrograph of a storm rainfall occurring on the upper part of the basin area has a longer time to peak than for a storm rainfall occurring on the lower basin area.

(6) **Time of concentration**: The time of concentration is the time needed for a drop of water falling on the most distant part of the river basin to reach the gauge station or the basin outlet (Fig. 2.30). It includes the time needed for all parts of the river basin to contribute run-off to the hydrograph and this time then indicates the highest discharge that can occur from certain storm intensity over the river basin area.

(7) **Lag time**: Lag time is the time distance between the centre of mass of effective rainfall and the centre of mass of the direct run-off hydrograph. As the determination of the centre of mass of the direct run-off hydrograph is difficult, lag time is also defined as the time distance between the centre of mass of effective rainfall and the peak of the direct run-off hydrograph (Fig. 2.30). It assumes uniform effective rainfall over the entire basin area.

Factors Affecting Hydrograph Characteristics

Several factors affect the characteristics of a stream hydrograph. These include:

(1) **Drainage characteristics**: The characteristics of the drainage are primarily derived from the parent geology of the drainage basin and affect the characteristics of the streamflow as well as the hydrograph. Large drainage basins receive more precipitation (rainfall) than the smaller basin, so have a greater peak discharge in comparison to smaller basins. Generally, smaller basins have shorter lag times than the larger basins because rainwater does not have to

travel long distances. Circular basins lead to shorter lag times and a greater peak discharge than elongated basins because the water in the former has a shorter distance to travel to reach a river. River basins with a steep slope are likely to have a shorter lag time than the basins with a gentle slope because the water in the former flows more quickly down to the river. Basins with high drainage density drain more quickly, so have a smaller lag time. In a saturated river basin, the surface run-off increases and rainwater comes into the river more quickly which reduces the lag time. If the drainage basin is dominated by impermeable rock, infiltration will be reduced and surface run-off will be higher which increases the peak discharge and reduces the lag time.

(2)   **Type and distribution of precipitation**: The precipitation in the form of snow is likely to have a greater lag time rather than rainfall because snow takes time to be melted before the water reaches the river channel. Amount of rainfall is very important to control the nature of the storm hydrograph. Heavy storm rainfall results in more supply of water in the drainage basin leading to a higher discharge of water.

(3)   **Soil and Land use**: Soil type and land use pattern may alter the characteristics of the hydrograph. Forest removal, grass cutting, urbanization, farming, building of roads and any other structures reduce the rate of infiltration and increases the run-off. The presence of more vegetation in the basin area intercepts precipitation (rainfall) and slows the draining of water into river channels and so the lag time increases.

(4)   **Human factors**: Rapid rate of urbanization by the human being increases the concentration of impermeable materials on the surface which reduces the infiltration level and surface run-off increases. This results in an increase in peak discharge and a shorter lag time.

Delineation of Run-Off Components in Storm Hydrograph

A streamflow hydrograph of a specific storm is a hydrograph of total run-off. Components of streamflow are (a) *direct run-off* and (b) *base flow*. Direct run-off is again divided into surface run-off and quick interflow, whereas the base flow is also divided into delayed interflow and groundwater run-off.

Surface Run-Off

Surface run-off is that portion of the run-off water that travels over the ground surface to the stream channel (Fig. 2.31a). Most surface run-off flows to the first-order channels because they collectively drain the largest area of the drainage basin. Surface run-off includes that portion of the precipitation directly falling on the water flowing in the channel but overland flow does not include this portion of precipitation.

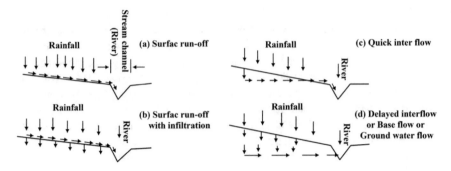

Fig. 2.31 Various components of run-off (after Singh 1994)

Interflow or Sub-surface Flow

It is the surface water that infiltrates the surface layer and moves laterally beneath the surface to a channel (Fig. 2.31c). Interflow can occur on forest floors, where the needles, leaves and other debris of the plants cover the ground surface. In interflow, the water is subject to higher flow resistance than the surface run-off. Because of this, the interflow water does not move as faster as surface run-off, hence delayed in reaching the stream channel.

Direct Run-Off

Direct run-off is considered to be the sum of surface run-off and the interflow. It is frequently equated with surface run-off. These two flow components move faster than groundwater flow and hence are often lumped together for hydrologic processes.

Base Flow

Base flow or groundwater flow is that component of the flow that contributed to the stream channel through groundwater (Fig. 2.31d). Groundwater occurs from surface water infiltration to the water table and then moving laterally to the stream channel through the aquifer. Such water moves very slowly than direct run-off and because of this reason it does not contribute to the peak discharge for a given storm hydrograph.

Important components of *streamflow* can be easily separated and illustrated in a *storm hydrograph*. In Fig. 2.32, point 'A' marks the initiation of the surface run-off, which is believed to end at the change in slope shown as 'B'; point 'B' is considered to be the initiation of interflow, which ends at point 'C'; point 'C' marks the initiation of groundwater flow, which continues up to the end of the hydrograph. Therefore, the segment of the curve from A to B indicates the contribution of surface run-off, B to C indicates the contribution of interflow, especially the quick interflow and beyond 'C' it indicates the contribution of delayed interflow or base flow or groundwater flow.

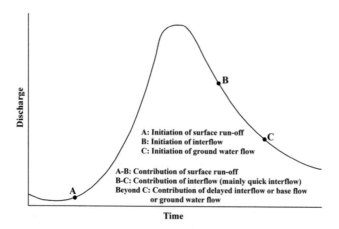

Fig. 2.32 Important components of streamflow hydrograph

2.5.4.6 Rating Curve

Rating curve (also called *stage–discharge relation curve*) is the graphical representation of the relationship between stream stage and stream flow or discharge of water (cusec or cumec) (Fig. 2.33 and Table 2.20) for a given point on a stream, generally at gauging stations. In rating curve, measured discharge is usually plotted on the *x*-axis (abscissa) and measured stage on the *y*-axis (ordinate) (Fig. 2.33). Stream stage (also called gauge height or stage) means the height of the water surface (in feet) above a well-known elevation where the stage is zero. Though the zero stage is arbitrary,

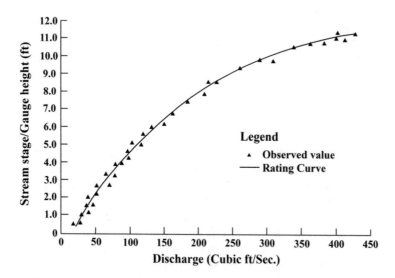

Fig. 2.33 Rating curve (Relationship between stream stage and discharge)

Table 2.20 Stream stage and discharge relationship

Sl. no	Gauge height (ft)	Discharge (cubic ft/s)	Sl. no	Gauge height (ft)	Discharge (cubic ft/s)
1	0.5	16	19	6.1.	150
2	0.65	26	20	6.8	162
3	1.0	30	21	7.25	163
4	1.15	37	22	7.6	210
5	1.5	35	23	8.6	220
6	2.0	38	24	8.5	230
7	2.2	51	25	9.3	269
8	2.6	52	26	9.6	315
9	3.1	78	27	9.5	296
10	3.25	62	28	9.75	310
11	3.85	78	29	10.4	340
12	4.1	95	30	10.6	371
13	4.5	96	31	10.7	367
14	4.65	104	32	10.75	384
15	4.9	120	33	10.9	401
16	5.0	103	34	10.95	406
17	5.6	122	35	11.1	430
18	5.8	132	36	11.2	407

generally it is close to the streambed. Stream discharge is measured several times over a range of stream stages. These measurements are done over a period of months or years for the establishment of an accurate relationship between the discharge and gauge height at the gauging station. Additionally, these relationships must constantly be verified against ongoing stream flow or discharge measurements because stream channels are always changing.

Controls of Rating Curve

Rating curve represents the integrated result of a wide variety of channel and flow parameters. The combined effect of these parameters is considered as control. If the rating curve (stage–discharge relationship) remains the same with time, then it is called permanent control. But if the relationship does change with time, it is called shifting control. When the control of a gauging station changes, the rating curve also changes. The change may be caused by (1) erosion or deposition, (2) rapidly changing flow, (3) varying backwater, and (4) changes in the flow because of dredging, channel encroachment and weed growth. For the shifting control due to cases (1) and (4), frequent current meter gauging is required (Singh 1994). Bedrock-bottomed parts of streams or metal/concrete structures or weirs are generally, though not always, thought of as permanent controls.

Steps of Development of Rating Curve

The development of rating curve generally involves three steps:

1. **Measuring stream stage**

A continuous record of the stream stage, i.e. the height of the water surface at a specific location along a stream is taken.

Various methods are used to measure the stream stage or gauge height. A common method is with a stilling well in the river bank or fixed to a bridge pier. Water from the stream comes into and leaves the stilling well through the underwater pipes which allow the water surface in the stilling well to be at the same height as the water surface in the stream. The height of the water surface inside the stilling well can be easily measured using pressure or float optic or acoustic sensor. The measured stream stage values are stored in an electronic data recorder device at a regular time interval, generally every 15 min interval.

A stilling well is not cost-effective to install; stream stage can also be determined by the measurement of the pressure needed to maintain a little flow of gas through a tube and bubbled out at a specified location below water in the stream. The pressure is directly linked to the height of the water column over the tube outlet in the stream. More the height of water above the tube outlet, more pressure is needed to push the gas bubbles through the tube.

2. **The discharge measurement**

Discharge of water (the volume of water passing a specific location along a stream per unit of time) is measured periodically. Generally, stream discharge is estimated by multiplying the area of stream cross-section by the average water velocity in that cross-section:

$$Discharge\ of\ water = area\ of\ the\ cross-section$$
$$\times\ average\ water\ velocity \qquad (2.16)$$

Numerous methods and types of equipment are used to measure the water velocity and cross-sectional area, including the water current meter and acoustic Doppler current profiler.

The current meter is used to measure the velocity of water at prefixed places (sub-sections) along a specified line, like a bridge or suspended cableway across a stream or river. In this technique, the stream cross-section is divided into several numbers of vertical sub-sections. The area of each sub-section is obtained by multiplying its width and depth, and the water velocity is measured using the current meter. Then the discharge of water in each sub-section is computed by multiplying the sub-section area and the measured water velocity. The total discharge in a cross-section is then calculated by summing the discharge of each sub-section.

Acoustic Doppler current profiler (ADCP) can also be used to measure the water discharge. An ADCP uses the Doppler effect to measure the water velocity. ADCP transmits a sound pulse into the stream water and detects the shift in the frequency of that pulse reflected back to the receiver of ADCP by sediment or other particulate matters transported in the water. The change in frequency or Doppler shift, which is detected by the ADCP, is then converted into water velocity. The discharge of water is then calculated by multiplying the cross-section area with the measured water velocity.

3. **The stage–discharge relation**

Identification of the natural and continuously changing relationship between the stream stage and discharge can be done by applying the stage–discharge relation to transform the frequently measured stream stage into estimates of discharge (USGS Science for a changing world).

Simple Rating Curve

The representation of measured values of stream stage and water discharge on an arithmetic scale results in a parabolic curve, which can be expressed as

$$Q = a(h - b)^c \tag{2.17}$$

where b is a constant indicating the gauge reading for zero discharge, a and c are rating curve constants. When the measured stream stage and water discharge data are represented on a logarithmic scale, a straight line results and the Eq. (2.17) becomes

$$\log Q = \log a + c \log(h - b) \tag{2.18}$$

The values of constants a and c can be obtained using the least square method. The value of constant b must be calculated beforehand and this can be done in different ways. A trial-and-error method can be used to get the value of b, which then gives the best-fit curve. Another way is to extrapolate the rating curve corresponding to $Q = 0$ and then plot log Q versus $\log(h - b)$. If the plotting of the values gives a straight line then the value of b obtained by extrapolation is correct and acceptable. Otherwise, another value in the vicinity of the previous value of b is chosen and the same procedure is repeated (Singh 1994). The value of b can also be computed analytically. From a smooth curve of Q versus h, the values of discharge like Q_1, Q_2 and Q_3 are selected in such a way that $\frac{Q_1}{Q_2} = \frac{Q_2}{Q_3}$. The corresponding values of the stage are h_1, h_2 and h_3. Then we have

$$\frac{(h_1 - b)^c}{(h_2 - b)^c} = \frac{(h_2 - b)^c}{(h_3 - b)^c} \tag{2.19}$$

or

$$\frac{h_1 - b}{h_2 - b} = \frac{h_2 - b}{h_3 - b} \tag{2.20}$$

from which the value of b is derived as

$$b = \frac{h_1 h_3 - h_2^2}{h_1 + h_3 - 2h_2} \tag{2.21}$$

Alternatively, the values of these parameters a, b and c can be obtained by optimization.

Generally, simple rating curve is satisfactory for most of the streams in which rapid fluctuations of stream stage are not experienced at the gauging section. The adequacy of the curve is measured by the scattering of values around the fitted curve. If there is a permanent control, the rating curve is basically permanent. In some gauging stations, there may be two or more controls each for a specific range of stream stage. The rating curve in such a station is discontinuous; the point of discontinuity corresponds to the stream stage revealing the change in control. An example is when submergence of a weir control starts when the tailwater level below the control rises above the lowest point of the control. Even in such situations, the simple rating curve may well be acceptable if the control is permanent, free of backwater and the slope of the streams is steep.

Uses of Rating Curve

Continuous measurement of stream gauges provide streamflow/discharge information which can be used for different purposes including.

(i)   Flood prediction
(ii)   Water management and allocation
(iii) Engineering design
(iv)   Research purposes
(v)   Operation of locks and dams
(vi)   Recreational safety and enjoyment etc.

2.5.4.7 Lorenz Curve and Gini Coefficient

The concentration or inequality in the distribution of any phenomenon or attribute or variable with respect to others can be studied in several different methods like (1) *Lorenz curve* and *Gini coefficient*, (2) *location quotient* and (3) *index of dissimilarity* (Mahmood 1999). The Lorenz curve is the graphical method and the Gini coefficient is the numerical or mathematical method of measurement of the degree of inequality of different phenomena or variables.

Lorenz curve (or Pareto curve), a form of percentage cumulative frequency curve, was first developed in 1905 by Max Otto Lorenz, an American economist for representing the inequality in the distribution of wealth or income. It is an effective graphical measure of inequality in the distribution of various items in social science, especially in geography, like studies on landholdings, income, expenditure, wealth, economic activities etc. (Pal 1998). It basically deals with the cumulative percentage distributions of the two attributes or variables at different points. For example, for the representation of income of the people in a country, this curve appears as a graph of population shares against their income shares, ordered from poorest to richest.

Techniques of Drawing of Lorenz Curve

(1) At first, both the variables are expressed in percentages (%), arranged according to ascending or descending order and their cumulative percentages are calculated (Tables 2.21, 2.22 and 2.23). Cumulative percentages of one variable (independent) are plotted on the x-axis and cumulative percentages of other variable (dependent) are plotted on the y-axis. For example, in the study of the number and area of landholdings, the cumulative percentage of the number of land holdings is plotted on the x-axis and the cumulative percentage of the area of landholdings is plotted on the y-axis (Fig. 2.34). Similarly, in the study of income distribution of the people, the cumulative percentage of population is plotted on the x-axis and the cumulative percentage of income is plotted on the y-axis (Fig. 2.36).

(2) Cumulative percentages of one variable up to certain points are plotted on a graph against the cumulative percentages of the other variable up to the same points. The different points so obtained are then joined by a smooth freehand curve, known as Lorenz curve.

(3) For comparison, a diagonal line at an angle of $45°$ is also drawn, joining the point of origin or lower-left corner ($x = 0\%$ and $y = 0\%$) and the last point or upper-right corner ($x = 100\%$ and $y = 100\%$) of the graph. This line is called 'line of equal distribution' (Figs. 2.34, 2.35 and 2.36). Lorenz curve will never cross the line of equal distribution.

How to Read the Lorenz Curve

(1) If 1% share of 'x' variable corresponds to 1% share of 'y' variable, 50% share of 'x' variable corresponds to 50% share of 'y' variable and n% share of 'x' variable corresponds to n% share of 'y' variable, then this is the condition of equal distribution of two variables. Thus the Lorenz curve in the ideal case would be a straight line (equal distribution line).

For example, if we want to understand the degree of inequality in the distribution of income of the people in a region or country and if 1% of the population has 1% of the total income, 50% of the population has 50% of the total income and n% population has n% of the total income, then the distribution of income among the people is perfectly equal.

(2) The deviation of any Lorenz curve from the equal distribution line is in proportion to the degree of inequality in the distribution of one variable in relation to

Table 2.21 Worksheet for Lorenz curve (The number and area of land holdings)

Size of land holdings (hectares)	No. of land holdings (in millions)	Area of land holdings (in million hectares)	% of no. of land holdings to total no. of land holdings	% of area of land holdings to total area of land holdings	Cumulative % of no. of land holdings to total no. of land holdings (X_i)	Cumulative % of area of land holdings to total area of land holdings (Y_i)	$X_i \cdot Y_{i+1}$	$Y_i \cdot X_{i+1}$
<2	28	48	26.17	6.64	26.17	6.64	456.14	310.29
2–4	22	78	20.56	10.79	46.73	17.43	1292.55	1107.68
4–6	18	74	16.82	10.23	63.55	27.66	2900.42	2171.31
6–10	16	130	14.95	17.98	78.5	45.64	5211.61	4051.92
10–15	11	150	10.28	20.75	88.78	66.39	7674.14	6390.70
15–20	8	145	7.48	20.05	96.26	86.44	9252.51	8462.75
20–25	3	70	2.80	9.68	99.06	96.12	9906	9612
>25	1	28	0.93	3.87	100	100		
Σ	107	723	100%	100%			36,693.37	32,106.65

Table 2.22 Worksheet for Lorenz curve (Total and urban population of six North Bengal districts of West Bengal)

Name of the districts	Total population (TP)	Urban population (UP)	% of UP to TP	% of TP to Grand TP (1)	% of UP to Total UP (2)	Ascending order of % of UP to TP	Arrangement of 1 (3)	Arrangement of 2 (4)
Darjeeling	1,842,034	718,175	38.99	10.71	22.43	10.25	16.41	9.03
Jalpaiguri	3,869,675	1,044,674	27.00	22.49	32.62	12.07	17.44	11.31
Koch Bihar	2,822,780	289,300	10.25	16.41	9.03	13.80	23.24	17.23
Malda	3,997,970	551,914	13.80	23.24	17.23	14.13	9.71	7.37
Uttar Dinajpur	3,000,849	362,187	12.07	17.44	11.31	27.00	22.49	32.62
Dakshin Dinajpur	1,670,931	236,075	14.13	9.71	7.37	38.99	10.71	22.43
Total	17,204,239	3,202,325		100%	100%		100%	100%

Cumulative % of 3 (X_i)	Cumulative % of 4 (Y_i)	$X_i \cdot Y_{i+1}$	$Y_i \cdot X_{i+1}$
16.41	9.03	333.78	305.66
33.85	20.34	1271.74	1161.21
57.09	37.57	2565.62	2509.68
66.8	44.94	5181.01	4012.69
89.29	77.56	8929	7756
100	100		
Σ		18,281.15	15,745.24

Source Census of India (2011)

Table 2.23 Inequality in the distribution of income of people of Sweden, USA and India

Decile	% Income of people			Cumulative % income of people		
	Sweden	USA	India	Sweden	USA	India
1	3.3	1.9	0.6	3.3	1.9	0.6
2	6.2	3.8	1.2	9.5	5.7	1.8
3	7.4	5.5	3.0	16.9	11.2	4.8
4	8.4	6.8	5.1	25.3	18.0	9.9
5	9.3	8.2	6.2	34.6	26.2	16.1
6	10.2	9.5	6.9	44.8	35.7	23
7	11.1	11.2	7.2	55.9	46.9	30.2
8	12.3	13.3	9.4	68.2	60.2	39.6
9	13.7	16.1	25.4	81.9	76.3	65.0
10	18.1	23.7	35.0	100	100	100

Sources Statistics Sweden, online database (2014), U.S. Census Bureau, Historical Income Tables (2016); Credit Suisse's Global Wealth Databook (2014)

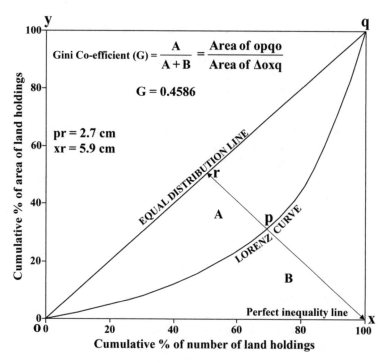

Fig. 2.34 Lorenz curve showing the inequality in the distribution of number and area of land holdings

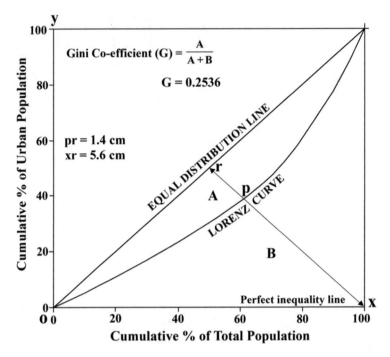

Fig. 2.35 Lorenz curve showing the inequality in the distribution of total and urban population

the other. Further, this Lorenz curve is from the line of equal distribution, so greater is the inequality. If the Lorenz curve coincides with the line of equal distribution, it indicates '0' inequality. But, if the curve coincides with '*x*' and '*y*' axes then it indicates the maximum (unity or 100%) inequality.

Gini Coefficient (G)

The inequality in the distribution of any phenomenon is numerically measured by an index known as *'Gini coefficient'*, developed by the Italian statistician Corrado Gini in the year 1912. Gini coefficient is the ratio of the area under the Lorenz curve and the equal distribution line to the area of the triangle formed by the *x*-axis, *y*-axis and the equal distribution line. Graphically, the Gini coefficient is defined as a ratio of two areas occupying the summation of all vertical differences between the Lorenz curve and the equal distribution line ('*A*' in Figs. 2.34 and 2.35) divided by the difference between the perfect equal distribution line and perfect inequality lines ('*A* + *B*' in Figs. 2.34 and 2.35). Therefore, the Gini coefficient can be defined as $\frac{A}{A+B}$ (shown in Figs. 2.34 and 2.35). In the case of uniform distribution, the Lorenz curve will fall on the line of equal distribution. Then the area between the curve and the line of equality would be zero ($A = 0$) and the value of Gini coefficient becomes 0 which means perfect equality. For the distribution with maximum concentration or

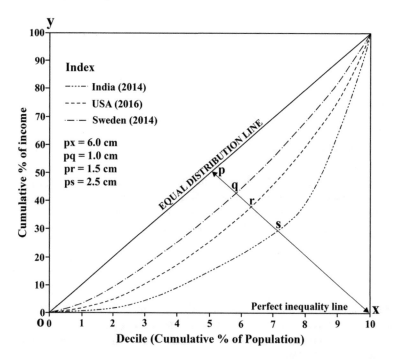

Fig. 2.36 Lorenz curve showing the inequality of income distribution of people in Sweden, USA and India *Sources* Statistics Sweden, online database (2014), U.S. Census Bureau, Historical Income Tables (2016); Credit Suisse's Global Wealth Databook (2014).

inequality, the curve will coincide with 'x' and 'y' axes. Then the area of 'B' would be 0 (B = 0), i.e. the area between the Lorenz curve and the equal distribution line becomes equal to the area of the triangle formed by the x-axis, y-axis and the equal distribution line and the value of Gini coefficient becomes unity or 1, which means perfect inequality. Thus the value of Gini coefficient varies on a scale between zero (0) and unity (1 or 100%) $[0 \leq G \leq 1]$.

The value of Gini coefficient (G) can be numerically calculated using the following formula:

$$G = \frac{1}{100 \times 100} \left| \sum (X_i . Y_{i+1}) - \sum (Y_i . X_{i+1}) \right| \tag{2.22}$$

where X_i and Y_i are the cumulative percentage distributions of the two attributes. For the data shown in Table 2.21,

$$G = \frac{1}{100 \times 100} |36693.37 - 32106.65|$$

$$G = \frac{1}{10000} |4586.72|$$

$$G = \frac{4586.72}{10000}$$

$$G = 0.4586$$

$$G = 0.46 \text{ (Round off)}$$

The value of *Gini coefficient* (G) can also be worked out graphically using the following technique:

In Fig. 2.34, the length of the straight line '*xr*' is 5.9 cm and this length is equivalent to the inequality of unity (1). The line '*xr*' cuts the Lorenz curve at the point '*p*' and the distance '*pr*' is 2.7 cm. Hence, the degree of inequality (G) can be calculated as $\frac{pr}{xr} = \frac{2.7 \text{ cm}}{5.9 \text{ cm}}$, i.e. $G = 0.4576 = 0.46$ (Round off).

Uses of the Lorenz Curve and Gini Coefficient

- Lorenz curve is the simplest representation and the Gini coefficient is the easiest measurement of inequality and can be interpreted easily.
- It is the most effective measure in comparing the differences between two and more data distributions.
- It can be easily applied to understand the change of distribution of any phenomenon within a country or region over a period of time, i.e. whether the inequality in the distribution is increasing or decreasing.
- It displays the distribution of wealth or income of a country or region among the population with the help of a graph.
- It can be effectively used while introducing specific measures for the development of the weaker sections in the economy.
- It can be applied to explain the fruitfulness of a government policy implemented to help the redistribution of income.

Problems of Using Lorenz Curve and Gini Coefficient

- Data restrictions, i.e. negative values cannot be represented.
- It might not always rigorously be accurate for a finite population.
- When two Lorenz curves intersect, it is difficult to ascertain which distribution illustrated by the curves represents more inequality.
- Gini coefficient as a measure of index of concentration should not be compared with the degree of concentration of a phenomenon or activity in a region to which the 'location quotient' is concerned with.
- Decision-makers and researchers are most interested in analysing inequality by a number.

For the data given in Table 2.22,

$$G = \frac{1}{100 \times 100}|18281.15 - 15745.24|$$

$$G = \frac{1}{10000}|2535.91|$$

$$G = \frac{2535.91}{10000}$$

$$G = 0.2536$$

$$G = 0.25 \text{ (Round off)}$$

In Fig. 2.35, the length of the straight line 'xr' is 5.6 cm and this length is equivalent to the inequality of unity (1). The line 'xr' cuts the Lorenz curve at the point 'p' and the distance 'pr' is 1.4 cm. Hence, the degree of inequality (G) can be calculated as $\frac{pr}{xr} = \frac{1.4 \text{ cm}}{5.6 \text{ cm}}$, or $G = 0.25$.

(a) **Inequality in the distribution of income of population in Sweden**

In Fig. 2.36, the length of the straight line 'px' is 6.0 cm and this length is equivalent to the inequality of unity (1). The line 'px' cuts the Lorenz curve of Sweden at the point 'q' and the distance 'pq' is 1.0 cm.

Hence, the degree of inequality (G) can be calculated as $\frac{pq}{px} = \frac{1.0 \text{ cm}}{6.0 \text{ cm}} = 0.17$.

(b) **Inequality in the distribution of income of population in the USA**

The line 'px' cuts the Lorenz curve of USA at the point 'r' and the distance 'pr' is 1.5 cm.

Hence, the degree of inequality (G) can be calculated as $\frac{pr}{px} = \frac{1.5 \text{ cm}}{6.0 \text{ cm}} = 0.25$.

(c) **Inequality in the distribution of income of population in India**

The line 'px' cuts the Lorenz curve of India at the point 's' and the distance 'ps' is 2.5 cm.

Hence, the degree of inequality (G) can be calculated as $\frac{ps}{px} = \frac{2.5 \text{ cm}}{6.0 \text{ cm}} = 0.42$.

Therefore, it is clear that the distribution of income of the people is more unequal in India ($G = 0.42$) compared to Sweden ($G = 0.17$) and the USA ($G = 0.25$). In India, the bottom 10% of the people possess only 0.6% of the total income of the country whereas it is 3.3% and 1.9% of total national income in the case of Sweden and the USA, respectively. The richest 10% of people in India enjoy 35% of the country's national income which is almost double the income of the top 10% of people in Sweden (18.1%). In the USA, the top 10% of people have 23.7% of the total national income. In India, 50% of the people have only 16.1% of the total income of the country whereas it is 34.6% and 26.2% of total national income in the case of Sweden and the USA, respectively.

2.5.4.8 Dispersion Graph

It is observed that in any set of data the actual values differ from each other and from the mean or average value also. The measurement and analysis of this spread-out character of the data set is called '*dispersion*'. In other words, dispersion indicates the degree of heterogeneity among the values in a data set. More the heterogeneity among the values, the more the degree of dispersion. Dispersion is as characteristic as the similarity is in statistics. It can be measured using two methods: (1) by means of the distances between specific observed values and (2) by means of the average deviations of individual observed values about the central value (Pal 1998).

When dispersion is measured in terms of the difference between the highest and the lowest values of the observations in a data set then it is called 'range'. It is used when the values in a data set form distance between points and individuals and when it is arranged graphically in terms of their magnitude, then a 'dispersion graph' is obtained. The total spread of the data within the range can be obtained from this graph. Symbolically speaking,

$$\text{Coefficient of range} = \frac{L - S}{L + S} \tag{2.23}$$

where L and S are the largest and smallest values, respectively.

Dispersion graphs are normally used to show the most important pattern in the distribution of the data set. The graph displays each value plotted as an individual point on a vertical scale. It shows the range of data and the distribution of each individual value within that range. Rainfall dispersion graph is one in which each year seasonal and annual amounts of rainfall are represented by placing a point against a vertical scale to enable to observe at a glance the span of dry years, normal years and wet years over a period of time.

Methods of Construction of Rainfall Dispersion Graph

- At first the fundamental values like median (Q_2), upper quartile (Q_3) and lower quartile (Q_1) etc. of the given data set are obtained using suitable formula.
- To obtain these values all the observations are arranged into ascending order. To obtain the value of the lower quartile (25% observations are smaller and 75% observations are larger than this value), consideration should be taken from the lower end and in the case of the upper quartile (75% observations are smaller and 25% observations are larger than this value) consideration should be taken from the upper end of the data set. In our given example with annual rainfall data for 30 years, the lower quartile (Q_1) will lie at the 7.75th position of the series, and the upper quartile (Q_3) will lie at 23.25th position (Table 2.24). The 15.5th value reckoned from the bottom or top of the graph indicates the median or Q_2 which divides the entire data set into two equal halves (Fig. 2.37).

Table 2.24 Calculations for rainfall dispersion graph (Annual rainfall of Bankura district, year 1976–2015)

Rainfall in ascending order	Position	Rainfall in ascending order	Position
1040	1	1620	16
1062	2	1645	17
1092	3	1678	18
1092	4	1690	19
1129	5	1765	20
1156	6	1780	21
1280	7	1828	22
1288	8	1856	23
1290	9	1856	24
1290	10	1876	25
1452	11	1959	26
1467	12	1975	27
1560	13	1993	28
1569	14	2014	29
1595	15	2128	30
Lower quartile (Q_1)	1286 mm (7.75th position)		
Median (Q_2)	1607.5 mm (15.5th position)		
Upper quartile (Q_3)	1856 mm (23.25th position)		
Co-efficient of Quartile Deviation	0.1814		

- A suitable vertical scale is selected and then each value of the data set (annual rainfall) are plotted graphically as individual points for the whole of the period under consideration on that vertical scale (Fig. 2.37).
- The central value (the median value, Q_2) is selected and this is displayed on the graph. Similarly the values of Q_1 and Q_3 are also displayed on the graph to understand the dispersion or variability among the observations within the graph.

Figure 2.37 illustrates the 30 years of annual rainfall distribution of Bankura district in a dispersion graph.

Advantages

- Easy to understand visually.
- Represents the spread of data from the mean and conveys much more information than other graphs drawn on mean values alone.
- Can find out the values of range, mean, median, mode, lower quartile, upper quartile and inter-quartile range.
- Can show the anomalies in the data set.
- It makes it possible to compare the variability of two or more sets of data.

Fig. 2.37 Rainfall
dispersion graph of Bankura
district (year 1976–2015)

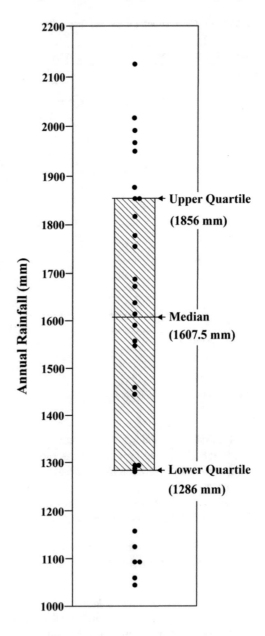

Disadvantages

- Better to work with lots of data.
- Sometimes the important features of the rainfall distribution may not be shown.

2.5.4.9 Rank-Size Graph

The *rank-size-rule* or *rank-size relationship* is an empirical and practical regularity of city size distribution observed in the urban systems in many countries in the world. It is an important method for analysing the total settlement networks in a region or country. Hence, it is a technique for understanding the national settlement system and facilitate the depiction and interpretation of the relationship between the population size and rank of the urban places. The rank-size-rule was first identified by Auerbach in 1913 but postulated and popularized by G.K. Zipf in 1949 in his book '*Human behaviour and the principle of least effort*'. After that, many geographers have studied the size distribution of settlements and described the relationship between the number and size of the settlements in geographical form.

In its general form, the rule states that, if all urban areas or cities in a country or region are ranked according to the population size with the largest city having the first rank, then the population of any urban area or city multiplied by its rank will equal the population of the first ranking city (largest city). In other words, the population of a city or urban area (P_r) of rank r can be calculated by dividing the population of the first ranking city (P_1) by its rank.

Symbolically, it can be written as

$$P_r = \frac{P_1}{r} \tag{2.24}$$

where P_r is the population of 'r' ranking city; P_1 is the population of the first ranking city and r is the rank of the city.

Accordingly, the second-ranking city of a country or region has half of the population of the largest city; the third-ranking city has one-third of the population of the largest city and so on down the scale (Table 2.25).

The relationship can be represented graphically by plotting the population of the city with respect to its rank. If rank (on 'x'-axis) and population (on 'y'-axis) are plotted using arithmetic scale then a curve (inverted J-shape) results (Fig. 2.38). But plotting of the population against the rank following logarithmic scale will produce a straight line (Fig. 2.39). When logarithmic scales are considered along both the axes then the equation can be rewritten as

$$\log P_r = \log\left(\frac{P_1}{r}\right) \tag{2.25}$$

$$\log P_r = \log P_1 - \log r \tag{2.26}$$

Rank-Size Graph According to Zipf (1949)

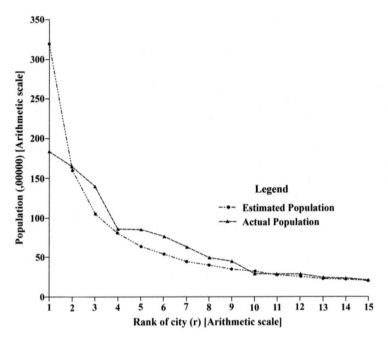

Fig. 2.38 Rank-size graph according to G.K. Zipf (arithmetic scale)

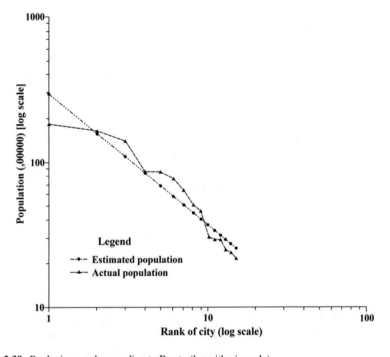

Fig. 2.39 Rank-size graph according to Pareto (logarithmic scale)

G.K. Zipf used the method, shown in Table 2.25 to compute the expected population of different cities or urban areas in a country or region.

Rank-Size Graph According to Pareto

According to this rule, the relation between the population of a town or city and its rank can be expressed as follows (Pareto's distribution):

$$P_r = K \cdot r^{-b} \tag{2.27}$$

where P_r is the population of the 'r' ranking city. K and b are the constants.

The above equation gets transformed into the following linear form after taking the logarithm on both sides:

$$\log P_r = \log(K \cdot r^{-b}) \tag{2.28}$$

$$\log P_r = \log K - b \, \log r \tag{2.29}$$

Table 2.25 Rank-size relationship of Indian cities (according to G.K. Zipf method)

City	Total (actual) population (2011 census)	Rank (r)	$\frac{1}{\text{rank } (r)}$	Expected or estimated total population $\left[\dfrac{\sum \text{Total population}}{\sum \frac{1}{r}}\right] \times \frac{1}{r}$
Mumbai	18,394,912	1	1	31,898,746
Delhi	16,349,831	2	0.5	15,949,373
Kolkata	14,035,959	3	0.33	10,526,586
Chennai	8,653,521	4	0.25	7,974,687
Bangalore	8,520,435	5	0.20	6,379,749
Hyderabad	7,674,689	6	0.17	5,422,787
Ahmedabad	6,361,084	7	0.14	4,465,824
Pune	5,057,709	8	0.125	3,987,343
Surat	4,591,246	9	0.11	3,508,862
Jaipur	3,046,163	10	0.10	3,189,875
Kanpur	2,920,496	11	0.09	2,870,887
Lucknow	2,902,920	12	0.08	2,551,900
Nagpur	2,497,870	13	0.077	2,456,203
Ghaziabad	2,375,820	14	0.07	2,232,912
Indore	2,170,295	15	0.067	2,137,216
Total	105,552,950		$\sum \frac{1}{\text{rank}} = 3.309$	

This equation can be equated with the regression equation:

$$Y = a - bX \qquad (2.30)$$

where $Y = \log P_r$; $X = \log r$; $a = \log K$.

$$b = \frac{\sum XY - \frac{\sum X \sum Y}{n}}{\sum X^2 - \frac{(\sum X)^2}{n}} \qquad (2.31)$$

$$b = \frac{80.1987 - \frac{12.11649 \times 101.0695}{15}}{11.40195 - \frac{(12.11649)^2}{15}}$$

$$b = \frac{80.1987 - \frac{1224.607586055}{15}}{11.40195 - \frac{146.8093299201}{15}}$$

$$b = \frac{80.1987 - 81.640505737}{11.40195 - 9.78728866134}$$

$$b = \frac{-1.441805737}{1.61466133866}$$

$$b = -0.8929462188$$

$$a = \overline{Y} - b\overline{X} \qquad (2.32)$$

$$a = \frac{101.0695}{15} - (-0.8929462188)\frac{12.11649}{15}$$

$$a = 6.73796666667 - (-0.8929462188) \times 0.807766$$

$$a = 6.73796666667 + 0.72129159538$$

$$a = 7.45925826205$$

Thus, $\log P_r = 7.45925826205 - 0.8929462188 \ \log r$.
Here, $a = \log K = 7.45925826205$ and $b = 0.8929462188$.
So, $\log K = 7.45925826205$.
where $K = $ Antilog of 7.45925826205.
Hence, $K = 28{,}791{,}100$.
So, the original equation (Eq. 2.27) can be written in the following form:

$$P_r = 28791100.r^{-0.8929462188} \qquad (2.33)$$

Table 2.26 Rank-size relationship of Indian cities (according to Pareto method)

City	Total (actual) population (P_r)	Rank (r)	$X = \log r$	X^2	$Y = \log P_r$	XY
Mumbai	18,394,912	1	0	0	7.2646977	0
Delhi	16,349,831	2	0.301029	0.090618	7.2135132	2.1714476
Kolkata	14,035,959	3	0.477121	0.227644	7.1472420	3.4100992
Chennai	8,653,521	4	0.602059	0.362475	6.9371928	4.1765993
Bangalore	8,520,435	5	0.698970	0.488559	6.9304617	4.8441848
Hyderabad	7,674,689	6	0.778151	0.605518	6.8850607	5.3576168
Ahmedabad	6,361,084	7	0.845098	0.714190	6.8035311	5.7496505
Pune	5,057,709	8	0.903089	0.815569	6.7039538	6.0542669
Surat	4,591,246	9	0.954242	0.910577	6.6619305	6.3570938
Jaipur	3,046,163	10	1	1	6.4837531	6.4837531
Kanpur	2,920,496	11	1.041392	1.084497	6.4654566	6.7330747
Lucknow	2,902,920	12	1.079181	1.164631	6.4628350	6.9745687
Nagpur	2,497,870	13	1.113943	1.240869	6.3975698	7.1265280
Ghaziabad	2,375,820	14	1.146128	1.313609	6.3758135	7.3074983
Indore	2,170,295	15	1.176091	1.383190	6.3365187	7.4523226
Total	$\sum P_r =$ 10,55,52,950	$\sum r =$ 120	$\sum X =$ 12.11649	$\sum X^2 =$ 11.40195	$\sum Y =$ 101.0695	$\sum XY =$ 80.1987

As per the rank-size relationship, by substituting 'r' $= 1, 2, 3, 4$ etc. in Eq. (2.33), we get the estimated population of cities ranking 1st, 2nd, 3rd, 4th etc. The population of the top 15 cities in India are estimated based on the fitted rank-size relationship in the given Eq. (2.33) and the results are given in Tables 2.26 and 2.27. The graphical representation of the result is shown in Fig. 2.39.

It must be noted that the differences between actual and estimated populations (shown in Table 2.27) have been calculated based on the relationship between 15 cities only. These differences may be reduced if more numbers of cities are taken into consideration in determining the relationship.

Types of Deviations in Rank-Size Rule

Three main types of deviations in the rank-size rule are as follows (Siddhartha and Mukherjee 2002):

Primary Deviation

The population of the second-largest city is less than half the population of the largest city, i.e. a condition for the development of *primate city* (the city having

Table 2.27 Expected populations and their deviations from actual populations

City	Rank (r)	Actual population	Estimated population	Difference	% Difference
Mumbai	1	18,394,912	28,791,100	1,03,96,188	36.11
Delhi	2	16,349,831	15,504,389	−845,442	−5.45
Kolkata	3	14,035,959	10,794,800	−3,241,159	−30.02
Chennai	4	8,653,521	8,349,319	−304,202	−3.64
Bangalore	5	8,520,435	6,840,937	−1,679,498	−24.55
Hyderabad	6	7,674,689	5,813,143	−1,861,546	−32.02
Ahmedabad	7	6,361,084	5,065,603	−1,295,481	−25.57
Pune	8	5,057,709	4,496,219	−561,490	−12.49
Surat	9	4,591,246	4,047,352	−543,894	−13.44
Jaipur	10	3,046,163	3,683,935	637,772	17.31
Kanpur	11	2,920,496	3,383,378	462,882	13.68
Lucknow	12	2,902,920	3,130,454	227,534	7.27
Nagpur	13	2,497,870	2,914,518	416,648	14.29
Ghaziabad	14	2,375,820	2,727,894	352,074	12.91
Indore	15	2,170,295	2,564,909	394,614	15.38
Total	$\sum r = 120$	$\sum P_r =$ 10,55,52,950	108,107,950	25,55,000	−30.23

twice or more population than the next ranking city in the urban hierarchy) emerges (Fig. 2.40). A primate city is developed when few simple strong forces operate rather than many complex forces operating. The small size of the country, long colonial history, simple economic and political organization, dual economy etc. are the factors leading to the development of city primacy. Bangkok (Thailand), Lagos (Nigeria), Harare (Zimbabwe) etc. are some examples of the primate city.

Binary Deviation

Binary deviation emerges when the population of the second-largest city is more than half the population of the largest city. This situation is observed when a number of cities of almost similar size dominate the upper end of the hierarchy (Fig. 2.40). The high rate of industrialization, presence of more than one national identity, long history of urbanization etc. are the factors responsible for binary deviation in the rank-size relationship. For example, Madrid and Barcelona in Spain; Mumbai, Delhi and Kolkata in India.

Stepped Pattern Deviation

In *stepped pattern deviation,* not one but a number of cities may be observed at every level, each city resembling the others in population size and functioning (Fig. 2.40).

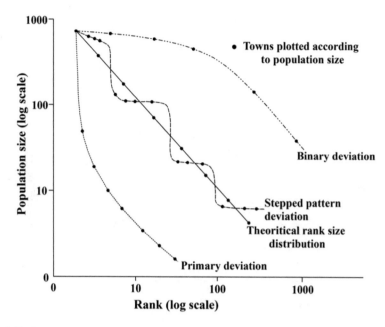

Fig. 2.40 Deviations in rank-size distribution

2.5.4.10 Box Plot ('Box-And-Whiskers') Graphs

The concept of *box-and-whiskers graph* was first given by John Tukey in 1970. A box plot, also known as box-and-whisker plot, is an important graphical method to represent the spread and centres of a data set. Measures of spread consist of the inter-quartile range and the mean, whereas the measures of the centre include the average or mean and median (the middle-most value) of a data set. Box-and-whisker plot is a data display that allows seeing many attributes of a distribution at a glance, i.e. they can be a useful means for getting a quick summary of the data set.

Elements of a Box-And-Whisker Plot

Box plot is a convenient method for the graphical depiction of groups of numerical data using five number summaries: the minimum, the maximum, the median, the lower quartile and the upper quartile (Figs. 2.41 and 2.42).

1. **Minimum**: It is the lowest value in the data set excluding outliers, if any and shown at the far left of the plot, i.e. at the end of the left 'whisker'.
2. **Maximum**: It is the largest value in the data set excluding outliers, if any and shown at the far right of the right 'whisker'.
3. **Median (Q_2)**: It is the middle-most value of the data set and is represented as a line at the centre (middle) of the box.

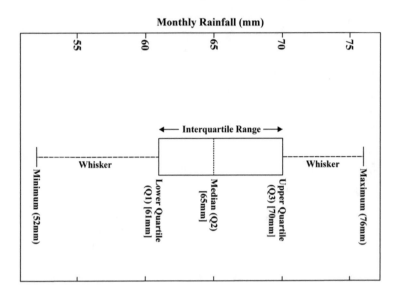

Fig. 2.41 Box-and-whisker graph without outliers

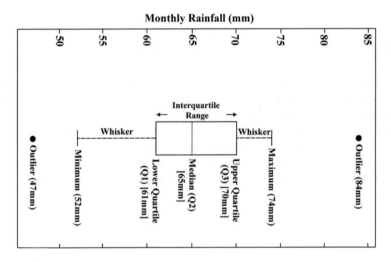

Fig. 2.42 Box-and-whisker graph with outliers

4. **First (lower) quartile (Q_1):** It is the middle value between the smallest number (not always the minimum) and the median (Q_2) of the data set and is shown at the far left of the box, i.e. at the far right of the left 'whisker'.

5. **Third (upper) quartile (Q_3):** It is the middle value between the largest number (not always the maximum) and the median (Q_2) of the data set and is shown at the far right of the box, i.e. at the far left of the right 'whisker'.

6. **Inter-quartile range (IQR)**: It is the distance between the lower and upper quartile.

$$IQR = Q_3 - Q_1 = q_n(0.75) - q_n(0.25) \tag{2.34}$$

Methods of Construction

In constructing this graph, at first we draw an equal interval scale and using this scale, a rectangular box is drawn with one end at the lower quartile (Q_1) and the other end at the upper quartile (Q_3). Then we draw a vertical line at the median value, i.e. at the second quartile (Q_2). A distance of 1.5 times the IQR is measured out from the right of the upper quartile and a horizontal line is drawn up to the larger observed value from the given data set that falls within this distance. In the same way, a distance of 1.5 times the IQR is measured out from the left of the lower quartile and a horizontal line is drawn up to the lower observed value from the data set that falls within this distance. These two horizontal lines or segments are called the '*whiskers*' (Figs. 2.41 and 2.42). All other observed values are plotted as outliers.

The spacing between different divisions of the box specifies the amount of dispersion or spread (degree of heterogeneity) and skewness (degree of symmetry or asymmetry) in the data set, and gives an idea about outliers.

Example Without Outliers

The monthly rainfall of 24 months was measured in mm and the values are given below:

52, 52, 52, 53, 58, 61, 61, 62, 62, 63, 64, 65, 65, 65, 66, 67, 68, 70, 70, 71, 72, 73, 74 and 76.

A box-and-whisker plot can be constructed by calculating the five number summaries: minimum, maximum, median, lower quartile and upper quartile.

The minimum is the smallest number in the given rainfall data set. Here, the minimum monthly rainfall is 52 mm.

The maximum is the largest number in the given rainfall data set. Here, the maximum monthly rainfall is 76 mm.

The median is the middle value of the ordered rainfall data set. This means that exactly 50% of the rainfall values are less than the median and 50% of the rainfall values are greater than the median rainfall. So, the median rainfall of the given data set is 65 mm.

The lower quartile is a value in which exactly 25% of the values are less than this and 75% of the values are greater than this value. It can be easily calculated by finding the middle value between the minimum value and the median value. In this given data set, the value of the lower quartile (middle value between 52 and 65 mm) is 61 mm.

The upper quartile is a value in which exactly 75% of the values are less than this and 25% of the values are greater than this value. It can be easily calculated by finding the middle value between the median value and the maximum value. In this given data set, the value of the upper quartile (middle value between 65 and 76 mm) is 70 mm.

The inter-quartile range (IQR) can be calculated using Eq. 2.34:

$$\mathbf{IQR} = Q_3 - Q_1 = \mathbf{70\ mm} - \mathbf{61\ mm} = \mathbf{9\ mm}$$

Hence, **1.5 IQR 1.5 IQR = 1.5 × 9 mm = 13.5 mm**
1.5 IQR after (above) the upper quartile is

$$Q_3 + \mathbf{1.5\ IQR} = \mathbf{70\ mm} + \mathbf{13.5\ mm} = \mathbf{83.5\ mm}$$

1.5 IQR before (below) the lower quartile is

$$Q_1 - \mathbf{1.5\ IQR} = \mathbf{61\ mm} - \mathbf{13.5\ mm} = \mathbf{47.5\ mm}.$$

Here, the maximum rainfall in the data set is 76 mm and 1.5 IQR after (above) the upper quartile is 83.5 mm which indicates that the largest data set value is lower than 1.5 IQR after (above) the upper quartile. Therefore, the upper whisker will be drawn at the maximum rainfall value, i.e. 76 mm.

In the same way, the minimum rainfall in the data set is 52 mm and 1.5 IQR before (below) the lower quartile is 47.5 mm, which indicates that the smallest data set value is greater than 1.5 IQR before (below) the lower quartile. Therefore, the lower whisker will be drawn at the minimum rainfall value, i.e. 52 mm (Fig. 2.41).

Example with Outliers

The above example is without outliers. But in this example outliers are incorporated by changing the first and last values of rainfall (in mm).

47, 52, 52, 53, 58, 61, 61, 62, 62, 63, 64, 65, 65, 65, 66, 67, 68, 70, 70, 71, 72, 73, 74 and 84.

As all the values except the first and last values remain unchanged, the median, lower quartile and upper quartile remain the same.

In this data set, the maximum rainfall value is 84 mm and 1.5 IQR after (above) the upper quartile is 83.5 mm, which indicates that the maximum rainfall is larger than 83.5 mm. So, the maximum rainfall value (84 mm) is an outlier. Therefore, the upper whisker will be drawn at the greatest rainfall value smaller than 83.5 mm, which is 74 mm.

In the same way, the minimum rainfall value is 47 mm and 1.5 IQR before (below) the lower quartile is 47.5 mm, which indicates that the minimum rainfall is smaller than 47.5 mm. So, the minimum rainfall value (47 mm) is an outlier. Therefore, the lower whisker will be drawn at the smallest rainfall value larger than 47.5 mm, which is 52 mm (Fig. 2.42).

2.5.4.11 Hypsometric Curve or Graph

Hypsometric (also called *hypsographic*) *curve* or graph is an important form of cumulative frequency curve or an Ogive. It is obtained by plotting the height of the contour with respect to the corresponding proportions of a specified unit area of the earth's surface (say a drainage basin) (Pal 1998). Hypsometry, first described by Strahler (1952), involves the measurement and analysis of the relationship between height and basin area to understand the degree of dissection and stage of the cycle of erosion. The basic data required for the study of area–height relationship are areas between successive contours and their respective heights. The area may be measured with the help of planimeter or may be estimated by the intercept method. The height is obtained from the contour map.

Area–height graph indicate actual areas between two successive contours and therefore the horizontal axis represents the area in terms of percentage of total area and the vertical axis represents the height. Hypsometric graph is generally used to show the proportion of the area of the surface at various elevations above or below a datum and thus the values of the area are plotted as ratios of the total area of the basin against the corresponding heights of the contours and hence the area is represented by cumulative proportion or percentage. A hypsometric curve is basically a graph representing the proportion of land area that exists at different heights by plotting relative area with respect to relative height.

On our earth, the heights can take on either positive (above sea level) or negative (below sea level) values and are bi-modal due to the contrast between the continents and oceans. Hypsometry of the earth reveals that earth has two peaks in height, one for the continents and the other for the ocean floors (Fig. 2.43).

From Table 2.28, it is clear that the total area of the whole basin (A) is 4830 km^2 and the maximum height of the basin (H) is 575 m. The area–height relationship of the basin (Fig. 2.44) can be expressed dimensionlessly by computing the relative area (the ratio between the individual area between successive contours, a_i and the whole area of the basin, A) and the relative height (the ratio of the mid-value of the contour height, h_i to the maximum height of the drainage basin, H). These ratios are computed in Table 2.28 and represented on graph as shown in Figs. 2.45 and 2.46. Although the hypsometric curve representing the relation between the proportions of area (shown on x-axis or abscissa) and height (shown on y-axis or ordinate) essentially pass through $X = 0$, $Y = 1$ and $X = 1$, $Y = 0$ but its location on graph is a function of the stage of erosion of the basin concerned.

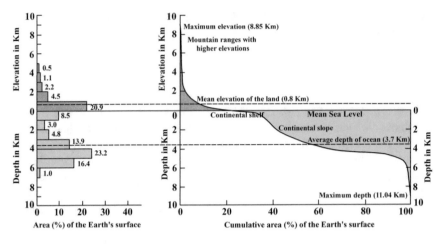

Fig. 2.43 Hypsometric curve for the whole earth

Table 2.28 Calculations for area–height graph and hypsometric curve in a sample drainage basin (Fig. 2.44)

Class intervals for height in metres	Mid value (h_i) in metres	Area between contours in sq. km.(a_i)	$\frac{h_i}{H}$	$\frac{a_i}{A}$	Cumulative up: less than $\frac{a_i}{A}$	$a_i h_i$
<200	175	130 (2.69%)	0.30	0.027	0.027	22,750
200–250	225	260 (5.38%)	0.39	0.054	0.081	58,500
250–300	275	680 (14.08%)	0.48	0.141	0.222	187,000
300–350	325	1230 (25.47%)	0.56	0.255	0.477	399,750
350–400	375	1080 (22.36%)	0.65	0.224	0.701	405,000
400–450	425	760 (15.73%)	0.74	0.156	0.857	323,000
450–500	475	420 (8.70%)	0.83	0.087	0.944	199,500
500–550	525	120 (2.48%)	0.91	0.025	0.969	63,000
> 550	Say 575·	150 (3.11%)	1.00	0.031	1.00	86,250
Total	$H = 575$	$\sum a_i = 4830$ km^2 $(= A)$				$\sum a_i h_i =$ 1,744,750

Hypsometric Integral (HI)

Hypsometric integral (HI) is the ratio of the volume or percentage of the total volume of the basin area below the curve (Fig. 2.46) and thus it indicates the volume of the basin area unconsumed by the dynamic wheels of erosion whereas *erosion integral (EI)* is a proportionate area above the curve and thus it reveals the volume of basin area which has already been consumed by the erosional processes. Thus hypsometric integral is the ratio between the area of the surface below the hypsometric curve (it

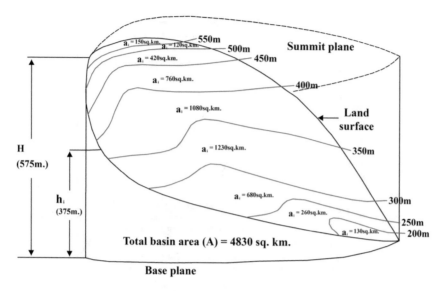

Fig. 2.44 Sample drainage basin showing height and area

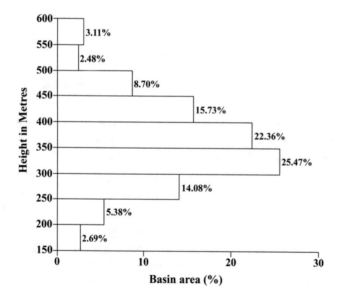

Fig. 2.45 Area–height relationship of the given drainage basin

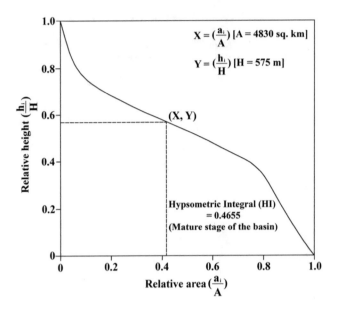

Fig. 2.46 Hypsometric curve of the given drainage basin

is between 0 and 1) and the area of the whole square (here it is 1). Theoretically, the value of hypsometric integral ranges between 0 and 1.

Though the value of hypsometric integral can be calculated using different techniques, the following techniques are very popular and widely accepted.

(1) **Elevation–relief ratio (E) relationship method**: When the spot heights of several numbers of places are known to us then the value of hypsometric integral is calculated using the following equation:

$$E \approx HI = \frac{El_{mean} - El_{min}}{El_{max} - El_{min}} \tag{2.35}$$

where E is the elevation–relief ratio equivalent to the hypsometric integral (HI); El_{mean} is the weighted mean elevation of the entire drainage basin; El_{max} and El_{min} are the maximum and minimum elevations of the drainage basin, respectively.

The weighted mean height of the drainage basin is calculated using the following formula:

$$\bar{h}_c = \frac{\sum a_i h_i}{\sum a_i} \tag{2.36}$$

where $\bar{h}_c$ is the mean height of the drainage basin. In the given example (Table 2.28 and Fig. 2.44), $\sum a_i h_i = 1,744,750$ and $\sum a_i = 4830$. So, the mean height of the drainage basin $(\bar{h}_c) = \frac{1744750}{4830} = 361.23$ m.

The value of El_{max} and El_{min} are 575 m and 175 m, respectively, then

$$E \approx HI = \frac{361.23 - 175}{575 - 175}$$
$$= \frac{186.23}{400}$$
$$= 0.4655$$

Hence, the value of hypsometric integral is 0.4655.

(2) When co-ordinates of the points define the hypsometric curve $(x, y;$ $x = \frac{a_i}{A}$ and $y = \frac{h_i}{H}$) (Fig. 2.46), HI is found using the following equation:

$$HI = \frac{\left| \sum x_i y_{i+1} - \sum y_i x_{i+1} \right|}{2} \tag{2.37}$$

(3) Mathematically, the hypsometric integral can be found from the integral calculus as integral

$$f = \frac{\text{Volume}}{\text{Total height} \times \text{Total area}} \int_{0.0}^{1.0} a.\Delta h \tag{2.38}$$

where a is area and Δh is the range in height h.

Importance of Hypsometric Curve and Hypsometric Integral

The value of *hypsometric integral* has been accepted as an important morphometric indicator of the stage of erosion of the basin. According to Strahler (1952), a high integral value exceeding 0.60 indicates the youthful stage in the development of drainage basin (denudation processes are not keeping pace with the rate of upliftment, i.e. much of the rock volume in the basin is still to be eroded); the value in between 0.35 and 0.60 indicates the mature or equilibrium stage and the value less than 0.35 indicates the old erosional surface or monadnock stage of the basin (Fig. 2.47).

The *hypsometric integral* is a dimensionless parameter and hence allows different drainage basins to be compared irrespective of scale. The shape of *hypsometric curve* and the value of hypsometric integral act as important indicators of basin conditions and characteristics. Hypsometric integral values are associated with the degree of disequilibrium in the balance between tectonic forces and the degree of erosion. Hypsometric integral is considered the most useful technique for the study of active tectonics.

A useful aspect of the hypsometric curve is that drainage basins having different sizes can be easily compared with each other since an area elevation is represented as

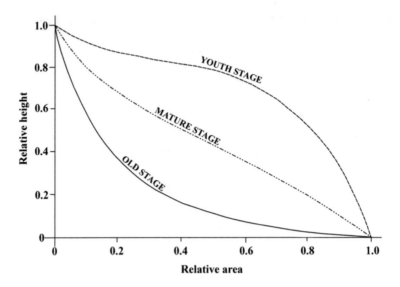

Fig. 2.47 Understanding the stages of landform development using hypsometric curve

functions of total area and total elevation, i.e. the hypsometric curve is independent of variations in basin size and relief (Alhamed and Ahmad Ali 2017).

It may be noted that the low value of hypsometric integral (below 0.30) is only sustained as long as few monadnocks give a relatively large difference in height between the highest and the lowest places. But when the monadnocks are eroded, the integral returns to about 0.40–0.60. It may be pointed out that hypsometric integral is a very delicate morphometric measure and hence it should be used with the greatest care and field verifications, otherwise it may render ambiguous results.

2.5.5 Frequency Distribution Graphs

2.5.5.1 Histogram

The most common and simple form of graphical representation of grouped frequency distribution is the *histogram*. It is constituted by a set of adjoining rectangles drawn on a horizontal baseline, having areas directly proportional to the class frequencies (Das 2009). Generally, the class boundaries are shown along the x-axis (abscissa) and the numbers of frequencies are represented along the y-axis (ordinate). As the class boundaries are taken into account to represent the rectangles, these become continuous to each other.

In constructing a histogram, the fundamental principle is that the area of each rectangle is directly proportional to the class frequency (f_i). Hence,

$$\text{Area of a rectangle } (A) \propto f_i \tag{2.39}$$

or

$$h_i \times w_i = k.f_i \qquad [\text{Area of a rectangle } (A) = \text{height} \times \text{width}] \tag{2.40}$$

where h_i is the height of the rectangle for the ith class; w_i is the width of the rectangle for the ith class; k is constant of proportionality.

or

$$h_i = \frac{k}{w_i} f_i \tag{2.41}$$

Equation 2.41 has distinctive applications on grouped frequency distribution with equal class size and unequal class size like (Sarkar 2015).

Grouped Frequency Distribution with Equal Class Size

In a grouped frequency distribution where all the classes are of equal size (w_i), Eq. 2.41 becomes

$$h_i = \left(\frac{k}{w}\right) f_i \tag{2.42}$$

or

$$h_i = k_i \times f_i; \quad [k_i = \frac{k}{w} = \text{constant because all the classes have equal size}]$$
$$\tag{2.43}$$

that is

$$h_i \propto f_i \tag{2.44}$$

So, in the case of the frequency distribution having the same class size, the height of each rectangle is directly proportional to its class frequency and it is then customary to take the heights numerically equal to the class frequencies (Table 2.29 and Fig. 2.48).

Grouped Frequency Distribution with Unequal Class Size

In a grouped frequency distribution where the classes are of different sizes (w_i), Eq. 2.41 becomes

Table 2.29 Grouped frequency distribution with equal class size (average concentration of SPM in the air)

Class boundary	Class mark (x_i)	Class width (w_i)	Frequency (f_i)
170.5–220.5	195.5	50	9
220.5–270.5	245.5	50	4
270.5–320.5	295.5	50	2
320.5–370.5	345.5	50	6
370.5–420.5	395.5	50	3
420.5–470.5	445.5	50	1

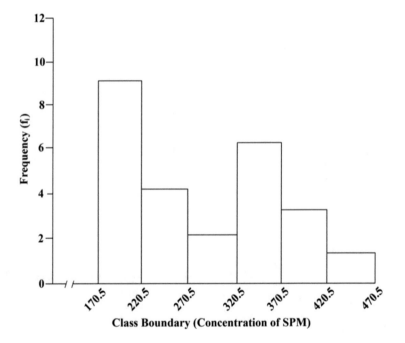

Fig. 2.48 Histogram (average concentration of SPM in the air)

$$h_i = k\left(\frac{f_i}{w_i}\right) \tag{2.45}$$

or

$$h_i = k \times f_{di} \quad [f_{di} \text{ is the frequency density of the } i\text{th class}] \tag{2.46}$$

that is,

$$h_i \propto f_{di} \tag{2.47}$$

Table 2.30 Grouped frequency distribution with unequal class size (monthly income of families)

Class boundary (Monthly income in Rs.'00)	Class mark (x_i)	Class width (w_i)	Frequency (f_i) [no. of family]	Frequency density (f_{di})$\left[\frac{f_i}{w_i}\right]$
0–50	25	50	40	0.8
50–120	85	70	60	0.86
120–250	185	130	45	0.35
250–350	300	100	35	0.35
350–600	475	250	25	0.1
600–950	775	350	20	0.057

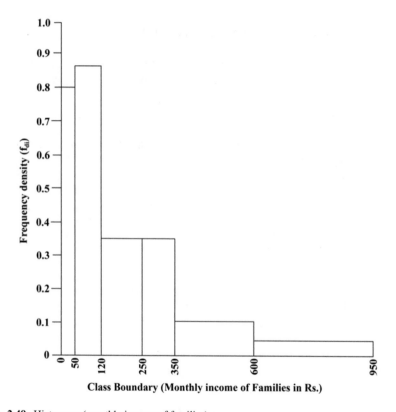

Fig. 2.49 Histogram (monthly income of families)

So, in the case of classes having unequal width, the rectangles will also be unequal in width and thus their heights must be directly proportional to the frequency densities but not to the class frequencies (Table 2.30 and Fig. 2.49). Therefore, in unequal class size, the rectangles of the histogram must be drawn with respect to the frequency densities of the classes.

Uses of Histogram

1. A series of rectangles or a histogram gives a visual description of the relative size of different groups of a data series. The entire distribution of total frequency into different classes becomes easy to understand at a glance.
2. The surface structure of the top of the rectangles gives an approximate idea about the nature (average, spread and shape etc.) of the frequency distribution and the frequency curve.
3. It is generally used for the graphical representation of mode.
4. Numerous geographical, economical and social data can be easily represented by histogram for their better and fruitful understanding.

2.5.5.2 Difference Between Historigram and Histogram

Historigram and *histogram* are two important methods of graphical representation of statistical or geographical data. The major differences between these two are as follows:

Historigram	Histogram
1. Representation of classified and summarized time series data by line is called historigram	1. Histogram is a set of adjoining rectangles drawn on a horizontal baseline
2. It represents the change of values of different variables with time	2. It represents the distribution of frequencies (number of observations) in different measurement classes
3. In historigram, time (year, month, day etc.) is shown along the x-axis and the values of the variable are shown along the y-axis	3. In histogram, the class boundaries are shown along the x-axis (abscissa) and the numbers of frequencies are shown along the y-axis (ordinate)
4. It is used to understand the temporal changes of uni-variate data and to compare the changes of two or more variables with time	4. It is used to understand the nature of the frequency distribution, drawing of frequency polygon and frequency curve, estimation of the value of mode etc.

2.5.5.3 Frequency Polygon

Frequency polygon is the graphical portrayal of grouped frequency distribution alternative to histogram and may be looked upon as it is derived from histogram by joining the mid-points of the tops of successive rectangles by straight lines. In construction, the frequency polygon is obtained by joining the consecutive points by straight lines whose abscissae indicate the class mark (x_i) and ordinates indicate the corresponding class frequencies (f_i) (Figs. 2.50, 2.51, 2.52 and 2.53).

Two main assumptions for constructing polygon are:

Fig. 2.50 Frequency polygon showing the average concentration of SPM in the air

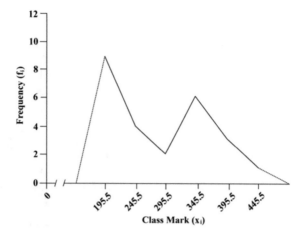

Fig. 2.51 Frequency polygon showing the monthly income of families

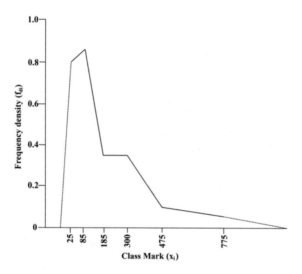

(i)   All the values in a particular class are uniformly distributed within the whole range of the class interval. Thus the class mark (x_i) is considered to be the representative of the corresponding class.

(ii)   For the same frequency distribution, the area covered by the histogram must be equivalent to the area enclosed within the frequency polygon. In this purpose, the two endpoints of the polygon are joined by straight lines to the abscissa at the mid values (class marks, x_i) of the empty classes at each end of the frequency distribution (Figs. 2.50, 2.51, 2.52 and 2.53).

Frequency polygon may be plotted separately and individually as well as on the histogram (Figs. 2.52 and 2.53). In the case of the distribution with unequal

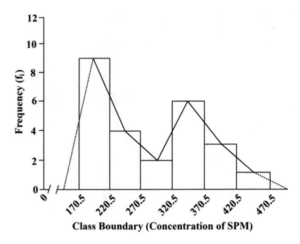

Fig. 2.52 Histogram with polygon showing the average concentration of SPM (mg/m^3) in the air

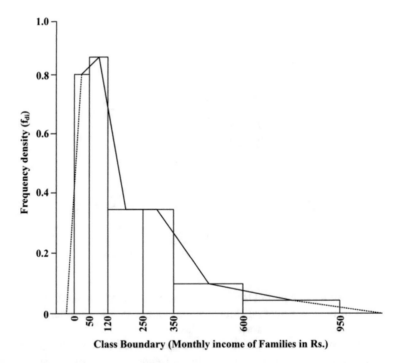

Fig. 2.53 Histogram with polygon showing the monthly income of families

classes, the polygon is drawn by plotting the frequency density (f_{di}) instead of simple frequency (f_i) against the class mark (x_i) (Figs. 2.51 and 2.53).

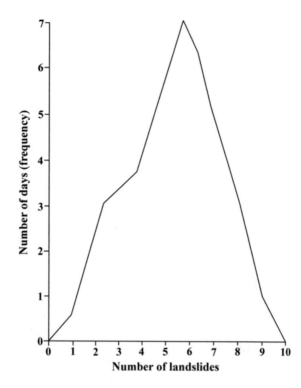

Fig. 2.54 Frequency polygon of discrete variable (Distribution of landslide occurrences)

Uses of Frequency Polygon

The frequency polygon is especially useful in representing simple frequency distribution of any discrete variable. For example, day-wise distribution of number of landslide occurrences can effectively be represented in a frequency polygon (Fig. 2.54). It gives us a better idea about the distribution of observations in different classes and the shape of the frequency curve.

2.5.5.4 Frequency Curve

In a generic sense, *frequency curve* is the modified form of histogram and frequency polygon. In drawing histograms, it is assumed that the observations (frequencies) are homogeneously distributed all through the range of values within the class boundaries of any class, but this may not be always true. Actually, a histogram provides the approximate idea about the nature and pattern of distribution of a limited number of frequencies (no. of observations) in different classes. The widths of the classes of the frequency distribution could be made smaller, but the problem is that some of the classes may remain empty (classes without any class frequency) and the actual pattern of the distribution of observations in the population will not be understood.

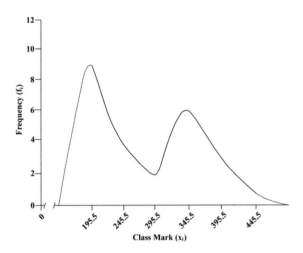

Fig. 2.55 Frequency curve showing the average concentration of SPM (mg/m³) in the air

But if the number of observations is very large, the situation will improve and all the classes are expected to have some number of frequencies, even when the widths of the classes are significantly small.

If the class width becomes smaller and smaller along with an indefinite increase of the total frequency, then the histogram and the frequency polygon tend to move towards a smooth curve called frequency curve (Das 2009). Generally, it is stated that in the frequency curve, the points obtained from the plotting of class frequency (f_i) or frequency density (f_{di}) against the class mark (x_i) are joined by a smooth curve instead of a series of straight lines (Figs. 2.55 and 2.56). Frequency curve represents the probability distribution of the variables in the population along with its area enclosed by the ordinate ('y'-axis) at two specified points on the abscissa ('x'-axis) indicating the probability of lying a value of the variable between these two extremes. Like histogram and frequency polygon, frequency curve is also an area graph.

Based on their shape and characteristics, frequency curve is of four types (Das 2009):

(i) Symmetrical bell-shaped or normal curve (Fig. 2.57a), (ii) Asymmetrical single-humped (Fig. 2.57b), (iii) J-shaped curve (Fig. 2.57c) and (iv) U-shaped curve (Fig. 2.57d).

Shape of the Frequency Curve

The *shape of the frequency curve* is very important as it represents the actual nature of a frequency distribution. It is generally measured in terms of two geometric properties of a frequency curve—(1) Symmetry or asymmetry (*skewness*) and (2) peakedness (*kurtosis*).

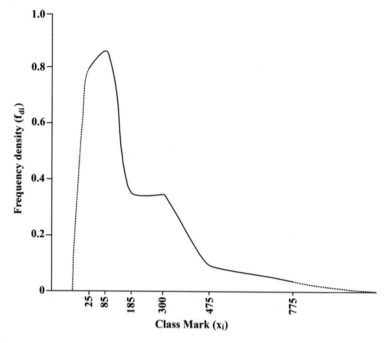

Fig. 2.56 Frequency curve showing the monthly income of families

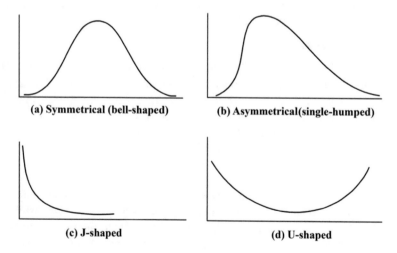

Fig. 2.57 Types of frequency curve

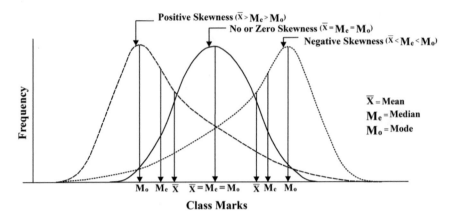

Fig. 2.58 Positive, negative and zero or no skewness

Skewness (S_k)

An important property of the shape of a frequency curve is whether it has one peak (unimodal) or more than one (bi-modal or multimodal). If it is unimodal (one peak), like most data sets, emphasis should be given to know whether it is symmetrical or asymmetrical in shape. *Skewness* measures the symmetry, or more accurately, the lack of symmetry of the frequency curve. Skewness signifies the extent of asymmetry of the frequency curve and is of three types:

(a)   **Positive skewness**

If most of the observations in a data set are located at the left of the curve (peak is towards the lower class boundaries) and the right tail is longer, then the distribution is said to be skewed right or *positively skewed*. In a positively skewed frequency distribution or curve, the relation between three measures of central tendency is *mean > median > mode*, i.e. the value of mean is greater than the median and again the value of median is greater than the mode (Fig. 2.58).

(b)   **Negative skewness**

If most of the observations in a data set are located at the right of the curve (peak is towards the upper-class boundaries) and the left tail is longer, then the distribution is said to be skewed left or *negatively skewed*. In a negatively skewed frequency distribution or curve, the relation between three measures of central tendency is *mean < median < mode*, i.e. the value of mean is lower than the median and again the value of median is lower than the mode (Fig. 2.58).

(c)   **Zero or no skewness or symmetric**

A frequency distribution or curve is said to be symmetrical in nature when the values are uniformly distributed around the mean. In such conditions the curve looks identical to the left and right of the central point. In a *symmetrical* or *zero skewed* frequency distribution or curve, the relation between three measures of central tendency is *mean* = *median* = *mode*, i.e. the values of mean, median and mode are equal (Fig. 2.58).

Skewness can be measured in different methods:

(1) Pearson's first measure

$$Skewness = \frac{Mean - Mode}{Standard\ deviation} \tag{2.48}$$

(2) Pearson's second measure:

$$Skewness = \frac{3(Mean - Median)}{Standard\ deviation} \tag{2.49}$$

(3) Bowley's measure:

$$Skewness = \frac{Q_3 - 2Q_2 + Q_1}{Q_3 - Q_1} \tag{2.50}$$

where Q_1, Q_2 and Q_3 are lower, middle and upper quartiles, respectively.

(4) Moment measure of skewness is called skewness coefficient, β_1 (read as beta-one):

$$Skewness\ coefficient(\beta_1) = \frac{\mu_3}{\sigma^3} \tag{2.51}$$

where μ_3 is the third central moment and σ is the standard deviation.

$$\mu_3 = \frac{\sum_{i=1,2....}^{n}(x_i - \overline{x})^3}{N}\ \text{(for ungrouped data)} \tag{2.52}$$

$$\mu_3 = \frac{\sum_{i=1,2....}^{n} f_i(x_i - \overline{x})^3}{N}\text{(for grouped data)} \tag{2.53}$$

In statistics, 'moment' (μ) is the mean of the first power of the deviation, i.e. the spacing of the size class or individual item in the frequency distribution from the mean, adding them up and dividing them by the total size of the distribution. This is the first moment (μ_1) about the mean which is symbolically written as

$$\mu_1 = \frac{\sum |x_i - \overline{x}|}{N} \tag{2.54}$$

Higher moments (μ_2, μ_3, μ_4 *etc.*) can be defined in the same way. In symmetrical frequency distribution both μ_1 and μ_2 are zero (0), so the skewness coefficient would also be zero (0).

According to Pearson's measures, there are no theoretical limits of skewness. But generally the value lies between $+3$ and -3. According to Bowley's measurement, the value of skewness ranges between $+1$ and -1.

Normal distribution (Normal Curve)

Normal distribution, also called Gaussian distribution (after the name of the mathematician Gauss), is a continuous probability distribution and is defined by the probability density function, $f(x)$ which is the height (Y) of the normal curve above the baseline at a given point (x_i) along the measurement scale of the random variable, x itself (Pal 1998). The model used to obtain the desired probabilities is

$$Y = f(x_i) = \frac{1}{\sigma_x \sqrt{2\pi}} e^{-\frac{1}{2}\left(\frac{x_i - \mu}{\sigma}\right)^2} \tag{2.55}$$

where e is the exponent (2.71828), π is the mathematical constant (3.14159), μ and σ are the population mean and standard deviation, respectively, x_i is any value of the continuous random variable $(-\infty < x_i < +\infty)$.

In other words, a frequency distribution having skewness $= 0$ $(S_k = 0)$ is called a normal probability distribution. The probability curve of the normal distribution is called *normal curve*. The curve is symmetrical and bell-shaped (Figs. 2.58 and 2.59) and the two tails extend to infinity on either side. In the real world, many actual distributions like rainfall data for any raingauge station collected over a large number of years (assuming no climatic change) tends to develop normal frequency distribution with a 'bell-shaped' symmetrical curve. This symmetry means that the height of the normal curve is the same if one moves equal distances to the left and right of the mean. The highest frequencies in this curve are around the mean and the frequency decreases as the distance from the mean increases. If we know the value of μ and σ, it is possible to determine the estimates of Y function for constructing the normal curve and calculating the area under the curve of any interval of x_i (Table 2.31).

Computation of the value of Y for every value of x_i is tedious and hence the position of any particular observation (say a score x_i) may be expressed relative to other scores in the data set by getting itself transformed to a standardized normal random variable known as *'standard score'* or a *'z-score'* (read as 'zee') where

$$z = \frac{x_i - \mu}{\sigma} = \frac{Critical\ value - Mean}{Standard\ deviation} \tag{2.56}$$

If all of the raw values in a distribution are converted to standard or z-scores, we get a new standardized distribution always having a 'mean, μ equal to zero (0)' and a 'standard deviation, σ equal to one (1)'. This is a useful transformation known as *standardization* which results in new values for the individuals. As the normal curve is symmetrical about the mean, half of the area under the curve lies on each side

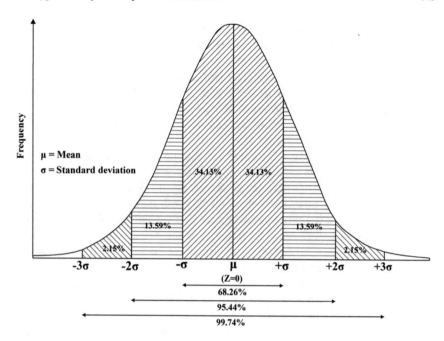

Fig. 2.59 Area under a standard normal curve

Table 2.31 Methods of calculating Y in $f(x)$ for constructing a normal curve

x_i (rainfall in cm) [$\mu = 50$ cm and $\sigma = 10$ cm]	$\frac{1}{\sigma\sqrt{2\pi}}$	$\frac{x_i-\mu}{\sigma}$	$\left(\frac{x_i-\mu}{\sigma}\right)^2$	$-\frac{1}{2}\left(\frac{x_i-\mu}{\sigma}\right)^2$	$e^{-\frac{1}{2}\left(\frac{x_i-\mu}{\sigma}\right)^2}$	Y (see Eq. 2.55)
25	0.03989	−2.50000	6.25000	−3.12500	0.04393	0.00175
45	0.03989	−0.50000	0.25000	−0.12500	0.88249	0.03520
50	0.03989	0.00000	0.00000	0.00000	1.00000	0.03989
65	0.03989	1.50000	2.25000	−1.12500	0.32465	0.01295
75	0.03989	2.50000	6.25000	−3.12500	0.04393	0. 00175

of the mean. This reveals that in a frequency distribution approximating the normal curve, 50% of the values will be less than the mean and 50% will be greater than the mean (Fig. 2.59).

Properties of Normal Curve

Normal curve has a number of interesting properties. These include:

(i)   Normal curve (normal distribution) has two important parameters μ and σ, μ = mean and σ = standard deviation. In some literature works, μ is also used as $\bar{x}$.

(ii)   The normal curve is always 'bell-shaped' and unimodal in nature.

(iii)   It is symmetrical about its centre (the line $x = \mu$) and mean occupies the centre. The vertical line drawn through the mean divides the curve into two equal halves. Thus, 50% of the values are less than the mean and 50% are greater than the mean.

(iv)   The values obtained from the addition or subtraction of the normally distributed values are also normally distributed.

(v)   Three measures of central tendencies are equal in value (i.e. mean = median = mode = μ), so on curve they coincide with each other.

(vi)   Normal curve has two points of inflections (the point where the curve changes curvature) at a distance of $\pm\sigma$ on either side of the mean (μ). Thus, the normal curve is convex upward in the interval ($\mu - 1\sigma$ and $\mu + 1\sigma$) and concave upward outside this interval.

(vii)   Two tails of a normal curve are asymptotic to the x-axis or horizontal axis, i.e. if two tails of a normal curve are extended in both the directions to infinity, they never cut the x-axis.

(viii) The form of the normal curve in terms of its shape (skewness) and height (kurtosis) depends on the mean and the standard deviation.

(ix)   The percentage distribution of the area under a standard normal curve is (Table 2.32 and Fig. 2.59):

    (a)   68.26% (68%) between $\mu \pm 1\sigma$
    (b)   95.44% (95%) between $\mu \pm 2\sigma$
    (c)   99.74% (99%) between $\mu \pm 3\sigma$

So, almost all the values of x will lie between the limits $\mu \pm 3\sigma$, i.e. mean $\pm$ 3(S · D.).

Kurtosis

In a frequency curve, it is required to know the 'convexity of the curve' which is *'kurtosis'* (Greek word meaning thereby 'bulkiness'). Two or more sets of data having an equal average, spread and symmetry but may differ in respect of their degree of peakedness. Kurtosis measures the degree of peakedness or convexity of a frequency curve, i.e. the extent to which values are concentrated in one part of the curve. It explains whether the distribution in the data set is having an excessively large or small number of values (observations) in the intermediate ranges between the mean and the extreme values and thus resulting in a peakedness or flat-toppedness of the frequency curve.

Kurtosis is the fourth moment about the mean, μ_4. According to Pearson, it is a coefficient, the kurtosis coefficient, β_2 (read as beta-two). Kurtosis coefficient (β_2)

Table 2.32 Standard normal distribution table

STANDARD NORMAL DISTRIBUTION (Values represent area to the left of the Z score)
(The left column stands for the first decimal value of z and the top row stands for the second decimal value of z)

z	0.00	0.01	0.02	0.03	0.04	0.05	0.06	0.07	0.08	0.09
0.0	0.5000	0.5040	0.5080	0.5120	0.5160	0.5199	0.5239	0.5279	0.5319	0.5359
0.1	0.5398	0.5438	0.5478	0.5517	0.5557	0.5596	0.5636	0.5675	0.5714	0.5753
0.2	0.5793	0.5832	0.5871	0.5910	0.5948	0.5987	0.6026	0.6064	0.6103	0.6141
0.3	0.6179	0.6217	0.6255	0.6293	0.6331	0.6368	0.6406	0.6443	0.6480	0.6517
0.4	0.6554	0.6591	0.6628	0.6664	0.6700	0.6736	0.6772	0.6808	0.6844	0.6879
0.5	0.6915	0.6950	0.6985	0.7019	0.7054	0.7088	0.7123	0.7157	0.7190	0.7224
0.6	0.7257	0.7291	0.7324	0.7357	0.7389	0.7422	0.7454	0.7486	0.7517	0.7549
0.7	0.7580	0.7611	0.7642	0.7673	0.7704	0.7734	0.7764	0.7794	0.7823	0.7852
0.8	0.7881	0.7910	0.7939	0.7967	0.7995	0.8023	0.8051	0.8078	0.8106	0.8133
0.9	0.8159	0.8186	0.8212	0.8238	0.8264	0.8289	0.8315	0.8340	0.8365	0.8389
1.0	0.8413	0.8438	0.8461	0.8485	0.8508	0.8531	0.8554	0.8577	0.8599	0.8621
1.1	0.8643	0.8665	0.8686	0.8708	0.8729	0.8749	0.8770	0.8790	0.8810	0.8830
1.2	0.8849	0.8869	0.8888	0.8907	0.8925	0.8944	0.8962	0.8980	0.8997	0.9015
1.3	0.9032	0.9049	0.9066	0.9082	0.9099	0.9115	0.9131	0.9147	0.9162	0.9177
1.4	0.9192	0.9207	0.9222	0.9236	0.9251	0.9265	0.9279	0.9292	0.9306	0.9319
1.5	0.9332	0.9345	0.9357	0.9370	0.9382	0.9394	0.9406	0.9418	0.9429	0.9441
1.6	0.9452	0.9463	0.9474	0.9484	0.9495	0.9505	0.9515	0.9525	0.9535	0.9545
1.7	0.9554	0.9564	0.9573	0.9582	0.9591	0.9599	0.9608	0.9616	0.9625	0.9633
1.8	0.9641	0.9649	0.9656	0.9664	0.9671	0.9678	0.9686	0.9693	0.9699	0.9706
1.9	0.9713	0.9719	0.9726	0.9732	0.9738	0.9744	0.9750	0.9756	0.9761	0.9767
2.0	0.9772	0.9778	0.9783	0.9788	0.9793	0.9798	0.9803	0.9808	0.9812	0.9817
2.1	0.9821	0.9826	0.9830	0.9834	0.9838	0.9842	0.9846	0.9850	0.9854	0.9857
2.2	0.9861	0.9864	0.9867	0.9871	0.9875	0.9878	0.9881	0.9884	0.9887	0.9890
2.3	0.9893	0.9896	0.9898	0.9901	0.9904	0.9906	0.9909	0.9911	0.9913	0.9916
2.4	0.9918	0.9920	0.9922	0.9925	0.9927	0.9929	0.9931	0.9932	0.9934	0.9936
2.5	0.9938	0.9940	0.9941	0.9943	0.9945	0.9946	0.9948	0.9949	0.9951	0.9952
2.6	0.9953	0.9955	0.9956	0.9957	0.9959	0.9960	0.9961	0.9962	0.9963	0.9964
2.7	0.9965	0.9966	0.9967	0.9968	0.9969	0.9970	0.9971	0.9972	0.9973	0.9974
2.8	0.9974	0.9975	0.9976	0.9977	0.9977	0.9978	0.9979	0.9979	0.9980	0.9981
2.9	0.9981	0.9982	0.9982	0.9983	0.9984	0.9984	0.9985	0.9985	0.9986	0.9986
3.0	0.9987	0.9987	0.9987	0.9988	0.9988	0.9989	0.9989	0.9989	0.9990	0.9990
3.1	0.9990	0.9991	0.9991	0.9991	0.9992	0.9992	0.9992	0.9992	0.9993	0.9993
3.2	0.9993	0.9993	0.9994	0.9994	0.9994	0.9994	0.9994	0.9995	0.9995	0.9995
3.3	0.9995	0.9995	0.9995	0.9996	0.9996	0.9996	0.9996	0.9996	0.9996	0.9997

(continued)

Table 2.32 (continued)

STANDARD NORMAL DISTRIBUTION (Values represent area to the left of the Z score)
(The left column stands for the first decimal value of z and the top row stands for the second decimal value of z)

z	0.00	0.01	0.02	0.03	0.04	0.05	0.06	0.07	0.08	0.09
3.4	0.9997	0.9997	0.9997	0.9997	0.9997	0.9997	0.9997	0.9997	0.9997	0.9998
3.5	0.9998	0.9998	0.9998	0.9998	0.9998	0.9998	0.9998	0.9998	0.9998	0.9998
3.6	0.9998	0.9998	0.9999	0.9999	0.9999	0.9999	0.9999	0.9999	0.9999	0.9999
3.7	0.9999	0.9999	0.9999	0.9999	0.9999	0.9999	0.9999	0.9999	0.9999	0.9999
3.8	0.9999	0.9999	0.9999	0.9999	0.9999	0.9999	0.9999	0.9999	0.9999	0.9999
3.9	1.0000	1.0000	1.0000	1.0000	1.0000	1.0000	1.0000	1.0000	1.0000	1.0000

is obtained using the following formula:

$$\beta_2 = \frac{\mu_4}{\mu_2^2} \tag{2.57}$$

Simply it can be written as

$$\beta_2 = \frac{\mu_4}{\sigma^4} \tag{2.58}$$

where μ_4 is the fourth central moment and σ is the standard deviation.

$$\mu_4 = \frac{\sum_{i=1,2....}^{n} (x_i - \bar{x})^4}{N} \quad \text{(for ungrouped data)} \tag{2.59}$$

$$\mu_4 = \frac{\sum_{i=1,2....}^{n} f_i (x_i - \bar{x})^4}{N} \quad \text{(for grouped data)} \tag{2.60}$$

Since kurtosis indicates the spread of the frequency curve, it is determined as

$$\text{Kurtosis measure, } K = \frac{Mean - Median}{Standard\ deviation} \tag{2.61}$$

For a symmetrically distributed curve, as the mean and median coincides with each other, the kurtosis measure, K should be zero and the kurtosis coefficient, β_2 should be 3.0, i.e. $K = \beta_2 - 3$, where with $K = 0$, β_2 becomes 3.0. Based on the peakedness of the frequency curve, three types of curves are identified:

(1) **Leptokurtic (Positive kurtosis):** When $\beta_2 > 3$ and $K > 0$, then the curve is more peaked than the perfect symmetrical curve and it is known as 'leptokurtic', i.e. the curve with a narrower central position and a higher tail than a perfect symmetrical curve (Fig. 2.60).

Fig. 2.60 Degree of
peakedness (Kurtosis) of
frequency curve

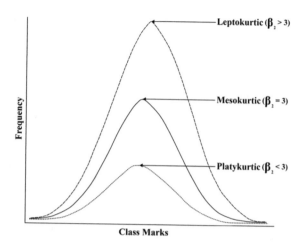

(2) **Platykurtic (Negative kurtosis)**: When $\beta_2 < 3$ and $K < 0$, then the curve is less peaked (more flat) and it is known as *'platykurtic'*, i.e. the curve with a broader central portion and a lower tail than the perfect symmetrical curve (Fig. 2.60).

(3) **Mesokurtic (Normal curve)**: From the kurtosis point of view, the perfect symmetrical curve is the curve having $\beta_2 = 3$ and $K = 0$ and it is known as *'mesokurtic'* (Fig. 2.60).

Uses of Frequency Curve

(i)   Different measures of central tendency and dispersion can be easily plotted on it.

(ii)   Plotting of frequency curves on the same base becomes effective to compare sets of data series.

(iii) The curve is also helpful to determine the normality of a data set.

2.5.5.5 Cumulative Frequency Polygon and Curve (Ogive)

The graphical representation of cumulative frequencies in a frequency distribution is called *cumulative frequency polygon*. For the drawing of this graph, the cumulative frequencies are plotted along the 'Y'-axis against the corresponding class boundaries, plotted along the 'X'-axis and the obtained points are joined by straight lines and this line is known as cumulative frequency polygon. Generally, two distinctive polygons are drawn: (i) Less than type and (ii) more than type.

(i)   **Less than type**: Less than type polygon is drawn based on the less than type cumulative frequencies (Tables 2.33 and 2.34). It begins from the lowest class boundary on the horizontal axis (abscissa or 'X'-axis), continues to rise upward

Table 2.33 Worksheet for drawing Ogive (with equal class size)

Class boundary	Frequency (f_i)	Cumulative frequency (F)			
		Less than	F	More than	F
170.5–220.5	9	170.5	0	170.5	25
220.5–270.5	4	220.5	9	220.5	16
270.5–320.5	2	270.5	13	270.5	12
320.5–370.5	6	320.5	15	320.5	10
370.5–420.5	3	370.5	21	370.5	4
420.5–470.5	1	420.5	24	420.5	1
	$N = \sum f_i = 25$	470.5	25	470.5	0

Table 2.34 Worksheet for drawing Ogive (with unequal class size)

Class boundary	Frequency (f_i)	Cumulative frequency (F)			
		Less than	F	More than	F
0–50	40	0	0	0	225
50–120	60	50	40	50	185
120–250	45	120	100	120	125
250–350	35	250	145	250	80
350–600	25	350	180	350	45
600–950	20	600	205	600	20
	$N = \sum f_i = 225$	950	225	950	0

and ends at the highest class boundary corresponding to the total frequency (N) of the distribution. The less than polygon looks like a broad and elongated S-shape.

(ii)   **More than type**: Contrary to less than type polygon, it is drawn based on more than type cumulative frequencies (Tables 2.33 and 2.34). It begins from the total frequency (N) at the lowest class boundary and progressively descends to the highest class boundary on the horizontal axis (abscissa or 'X'-axis). More than type polygon looks like a broad, elongated but inverted letter S.

Cumulative frequency curve is the modified form of cumulative frequency polygon in which the plotted points are joined by smooth freehand curves as an alternative of straight lines (Figs. 2.61 and 2.62). The combined representation of less than type and more than type cumulative frequency curves looks like a wine-glass called wine-glass curves. A combined representation of less than and more than cumulative frequency polygons or curves is called Ogive (Figs. 2.61 and 2.62). The two polygons or curves intersect at the median point of the distribution. The method of graphical construction of Ogives in frequency distribution with unequal class widths is the same as in the case of equal widths of the classes in frequency distribution.

Fig. 2.61 Cumulative
frequency curve (Ogive)
showing the average
concentration of SPM
(mg/m^3) in air

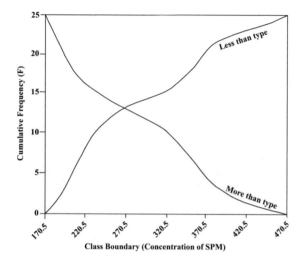

Fig. 2.62 Cumulative
frequency curve (Ogive)
showing the monthly income
of families

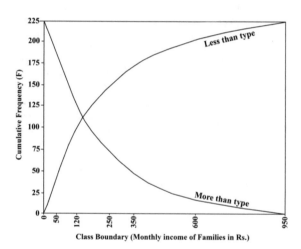

Uses of Cumulative Frequency Polygon and Curve (Ogive)

A cumulative frequency curve (Ogive) is more useful than a frequency curve to understand the content of a frequency distribution. The uses of Ogive include:

(1)   As the Ogive is the only graphical representation of the cumulative frequency distribution, it is very useful to find the values of median, quartiles, deciles and percentiles graphically.

(2)   The number of observations (frequencies) lying below or above a particular value, in between any two specified values can be easily found from the Ogive.

(3)   It is also useful to find out the cumulative frequencies above or below a certain specified value of the variable.

References

Alhamed M, Ahmad Ali S (2017) Hypsometric curve and hypsometric integral analysis of the Abdan Basin, Almahfid Basement Rock, Yemen. National seminar on recent advances and challenges in geochemistry, Env Sed Geol

Chow VT (1959) Open channel hydraulics. McGraw-Hill, New York

Das NG (2009) Statistical methods (Volume I & II). McGraw Hill Education (India) Pvt Ltd ISBN: 978-0-07-008327-1

Geddes A, Ogilvie AG (1938) The technique of regional geography. Jour of MGS 13(2):121–132

Mahmood A (1999) statistical methods in geographical studies. Rajesh Publication. ISBN: 9788185891170, 81-85891-17-6

Mitra A (1964) A functional classification of Indian towns. Institute of Economic Growth, India

Pal SK (1998) Statistics for geoscientists: techniques and applications. Concept Publishing Company, New Delhi, ISBN: 81-7022-712-1

Saksena RS (1981) A handbook of statistics. Indological Publishers & Booksellers

Sarkar A (2015) Practical geography: a systematic approach. Orient Blackswan Private Limited, Hyderabad, Telengana, India, ISBN: 978-81-250-5903-5

Siddhartha K, Mukherjee S (2002) Cities, Urbanisation and urban systems. Kisalaya Publications. ISBN: 81-87461-00-4

Singh VP (1994) Elementary hydrology. Prentice Hall of India Private Limited, New Delhi

Singh RL, Singh RPB (1991) Elements of practical geography. Kalyani Publishers

Sokolov AA, Chapman TG (eds) (1974) Methods for water balance computations. An international guide for research and practice, studies and reports in hydrology 17. UNESCO Press, Paris

Strahler A (1952) Dynamic basis of geomorphology. Geol Soc Am Bull 63:923–938. https://doi.org/10.1130/0016-7606(1952)63[923:DBOG]2.0.CO;2

Sutcliffe et al (1981) The water balance of the Betwa basin, India/Le bilan hydrologique du bassin versant de Betwa en Inde. Hydrol Sci Bull 26(2):149–158. https://doi.org/10.1080/026266681 09490872[J.V.SUTCLIFFE,R.P.AGRAWAL&JULIAM.TUCKER]

Taylor TG (1949) The control of settlement by humidity and temperature (with special reference to Australia and the Empire): an introduction to comparative climatology. Melbourne, VIC, Commonwealth Bureau of Meteorology

Zipf GK (1949) human behaviour and the principle of least effort, An introduction to human ecology. Addison-Wesle, Cambridge, MA

Chapter 3
Diagrammatic Representation of Geographical Data

Abstract Diagrammatic representation and visualization of geographical data is very simple, attractive and easy to understand and explain to the geographers as well as to the common literate people. It helps to explore the nature of data, the pattern of their spatial and temporal variations and understanding their relationships to accurately recognize and analyse features on or near the earth's surface. This chapter focuses on the detailed discussion of various types of diagrams classified on a different basis. All types of one-dimensional (bar, pyramid etc.), two-dimensional (circular, triangular, square etc.), three-dimensional (cube, sphere etc.) and other diagrams (pictograms and kite diagram) have been discussed with suitable examples in terms of their appropriate data structure, necessary numerical (geometrical) calculations, methods of construction, appropriate illustrations, and advantages and disadvantages of their use. It includes all the fundamental geometric principles and derivation of formulae used for the construction of these diagrams. A step-by-step and logical explanation of their construction methods becomes helpful for the readers for an easy and quick understanding of the essence of the diagrams. Each diagram represents a perfect co-relation between the theoretical knowledge of various geographical events and phenomena and their proper practical application with suitable examples.

Keywords Diagrammatic representation · Geometric principles · One-dimensional diagram · Two-dimensional diagram · Three-dimensional diagram

3.1 Concept of Diagram

Diagram is another important form of visual representation of geographical data in which importance is laid on the basic facts of one selected element. In the diagram, data are represented in a very much abstract and conventionalized geometric form. All types of categorical and geographical data, including time series and spatial series data, can be easily represented in diagrams. Representation of different geographical data by suitable diagrams is easy to understand and appreciated by all the people without having geographical, geometrical and statistical knowledge.

3.2 Advantages and Disadvantages of Data Representation in Diagrams

Advantages

The advantages of representation of data in diagram are:

(i)   Diagram makes the data simple, attractive and impressive, so easily intelligible to all.
(ii)   It saves a considerable amount of time, labour and energy.
(iii)   Comparison of two or more sets of data becomes possible and easy.
(iv)   It has universal utility, i.e. the technique is used all over the world.
(v)   It becomes helpful to detect the errors in data if any.
(vi)   Various complex data can be easily and simply represented by diagrams.
(vii)   It has an immense memorizing effect.

Disadvantages

In spite of all these advantages, representation of data by diagram has some limitations:

(i)   Diagram does not provide a detailed description of data
(ii)   Data can't be represented completely and accurately in a diagram.
(iii)   Portrayal of small variation in two or more sets of data is difficult in a diagram.
(iv)   Most of the diagrams are useful to the common people but these are of little significance to the professionals. Again, some of the diagrams (three-dimensional or multi-dimensional diagrams) are specifically useful to professionals and experts.
(v)   Depiction of three or more sets of data becomes difficult and impossible in diagram.
(vi)   Diagram does not represent the overall characters of data, rather it signifies the general conditions only.
(vii)   The application of the diagram is very limited in applied research.

3.3 Difference Between Graph and Diagram

Graphs and *diagrams* are two important techniques for the representation of statistical as well as geographical data, but major differences between them are as follows:

Graph	Diagram
1. Representation of series of data on graph paper either by means of Cartesian or polar or oblique co-ordinates on a reference frame is called graph	1. Representation of data in a highly abstract and conventionalized geometric form on two-dimensional plain paper is called diagram
2. Principles of co-ordinate geometry are applied immensely for the drawing of graphs	2. Some geometrical principles are applied but co-ordinate geometry is little or insignificantly applied in diagrams
3. Easy to draw, because it requires only the knowledge of co-ordinate geometry	3. Drawing of diagrams requires efficiency, experience and artistic knowledge. Lack of these will make the diagram less attractive and impressive
4. Graph depicts the functional or mathematical relationship between two or more variables	4. Diagram does not depict any functional or mathematical relationship between variables; it is used for comparisons only
5. Generally, time series data and frequency distribution data are appropriately represented in a graph	5. Diagrams are constructed for representing categorical data, including time series and spatial series data
6. Graphs are very much suitable and used for statistical analysis of geographical data	6. Diagrams are less suitable for the statistical analysis of geographical data
7. The value of median and mode can be easily estimated from a graph	7. Median and mode can't be estimated from a diagram
8. Graphs are less attractive and impressive to the eye	8. Diagrams are attractive to the eye and are better suited for publicity and propaganda
9. In the graph, data are represented by points or lines	9. In the diagram, data are represented by bars, pies, rectangles, squares, spheres etc.

3.4 Types of Diagrams in Data Representation

Based on the type and nature of data, the following categories of diagrams can be distinguished, i.e. statistical diagrams, geographical diagrams and statistical-geographical diagrams (Sarkar 2015). In addition to this, on the basis of the geometry of the figures to be constructed, diagrams may be classified into different types from which the geographers or researchers have to select the most suitable one (Table 3.1).

3.4.1 One-Dimensional Diagrams

It is the diagram in which the size of only one dimension, i.e. length, is considered to be fixed in proportion to the value of the data it represents.

Table 3.1 Types of diagrams

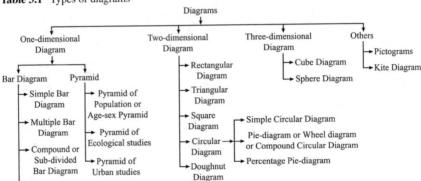

3.4.1.1 Bar Diagram

The representation of statistical or geographical data in the form of bars is called *bar diagram*. It consists of a number of bars that are equal in width and equally spaced. The bars are drawn on a common baseline on which the length or height of the bar is directly proportional to the value it signifies. Based on the constructional arrangements of the bars, three categories are identified: (i) *Vertical or columnar bar diagram*: Bars are drawn vertically above the abscissa (x-axis). (ii) *Horizontal bar diagram*: Bars are drawn parallel to the abscissa along the ordinate (y-axis). (iii) *Pyramidal bar diagram*: Horizontal bars are arranged in such a way that it forms a pyramid.

Principles of Construction of bar Diagrams

Though there is no hard and fast rule in constructing bar diagram but some important principles are followed for drawing it.

1. The bars should be neither too short nor too long. In other words, the bars should be proportionate in length and breadth.
2. The baseline from which bars are drawn should be clearly shown.
3. The scale should be mentioned clearly and accurately.
4. The intervening space between bars should be equal, i.e. bars should be drawn at an equal distance from each other.
5. The bars should be coloured or shaded in order to make them impressive and attractive.
6. Generally, vertical bars are used to represent the time series data whereas horizontal bars are used to depict the data classified geographically or data classified by their attributes.

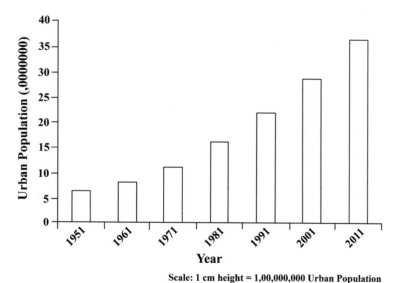

Scale: 1 cm height = 1,00,000,000 Urban Population

Fig. 3.1 Vertical simple bar (Temporal change of urban population in India since independence)
Source Census of India, 2011

Advantages and Disadvantages of the Use of bar Diagrams

The major advantages and disadvantages of the use of bar diagrams are as follows:

Advantages

1. Drawing and understanding of bar diagram is very simple and easy.
2. A large number of data can be easily represented in bar diagram.
3. Bar diagram can be drawn either vertically or horizontally.
4. It facilitates comparison of different data series.

Disadvantages

1. It is very difficult to represent a large number of aspects of any data in bar diagram.
2. The drawer fixes the width of the bars arbitrarily.

Types of Bar Diagrams

Simple Bar Diagram

The bar diagram showing only one component or category of data is called *simple bar diagram*. In this type each bar represents a single value only (Tables 3.2 and 3.3). It can be drawn either on a horizontal base (Fig. 3.1) or a vertical base (Fig. 3.2), but bars on a horizontal base are frequently used. The width of bars must be equal and they

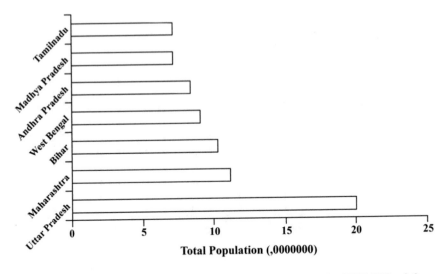

Fig. 3.2   Horizontal simple bar (Total population in selected states in India) *Source* Census of India, 2011

Table 3.2   Data for vertical simple bar diagram (Temporal changes of urban population in India)

Year	Urban population	Scale selection	Height of each bar (cm)
1951	62,443,709	1 cm to 1,00,000,000 urban population	0.62
1961	78,936,603		0.79
1971	109,113,977		1.09
1981	159,462,547		1.59
1991	217,177,625		2.17
2001	285,354,954		2.85
2011	361,986,870		3.62

Source Census of India, 2011

should be spaced with equal distance from one another. The scale for constructing simple bar diagram should be selected based on the highest and lowest values of the data to be represented.

Example Total population in different states of India, total population in different years or decades in India, year-wise production of wheat in India, state-wise production of rice in India, coal production in different countries in the world etc. can be represented by simple bar diagram.

Table 3.3 Data for horizontal simple bar diagram (Total population in selected states in India, 2011)

Name of the state	Total population (2011)	Scale selection	Length of each bar (cm)
Uttar Pradesh	199,581,477	1 cm to 50,000,000 population	3.99
Maharashtra	112,372,972		2.24
Bihar	103,804,673		2.08
West Bengal	91,347,736		1.82
Andhra Pradesh	84,665,533		1.70
Madhya Pradesh	72,383,628		1.44
Tamil Nadu	72,138,958		1.44

Source Census of India, 2011

An important limitation of simple bar diagrams is that they can represent only one component or one category of data. For example, while depicting the total population of different states in India, we can depict only the total population but sex-wise distribution of population in different states can't be represented in simple bar diagram.

Multiple Bar Diagram

Bar diagram in which different bars or proportionate lengths are drawn side by side representing the components is called *multiple bar diagram* (Fig. 3.3). It is generally used to compare two or more sets of statistical or geographical data (Table 3.4). In order to discriminate different bars, they are either differently coloured or different types of crossings or dottings are used in them. Multiple bar diagram is always equipped with an index or legend to indicate the meaning of different dotting or colours (Fig. 3.3).

Sub-Divided or Compound Bar Diagram

A *compound bar diagram* is one in which a single bar is sub-divided into different parts in proportion to the values given in the data (Fig. 3.4). When the data is composed of more than one component within a total then compound bar diagram is used (Table 3.5). The single bar represents the total or aggregate value while the component parts represent the component values of the aggregate. Sub-divisions in a bar are distinguished by using different colours or dottings or crossings. A legend or index is also given to indicate the meaning of different colours or dottings. This diagram reflects the relation among different components and also between different components and the aggregate. Compound bar diagram is also known as composite or component bar diagram.

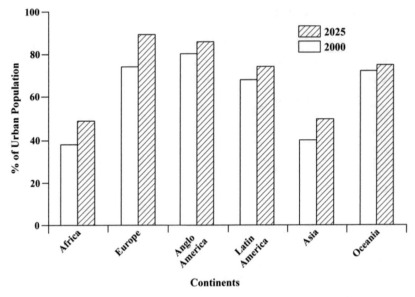

Scale: 1 cm height = 20% Urban Population

Fig. 3.3 Multiple bars showing the continent-wise urban population (%) in 2000 and 2025[*] *Source* UN Population Division, 2009–2010 and The World Guide, 12th ed. [*] Projected figures

Table 3.4 Calculations for multiple bar diagram (Continent-wise urban population)

Name of the continent	Percentage (%) of urban population		Scale selection	Height or length of each bar (cm)	
	2000	2025[*]		2000	2025[*]
Africa	38.5	49.6	1 cm to 20% urban population	1.92	2.48
Europe	75.0	90.0		3.75	4.5
Anglo America	80.0	86.0		4.0	4.3
Latin America	68.5	75.0		3.42	3.75
Asia	40.0	50.0		2.0	2.5
Oceania	72.5	75.3		3.62	3.76

[*] Projected figures
Source UN Population Division, 2009–2010 and The World Guide, 12th ed

Percentage Bar Diagram

Percentage bar diagram is a special form of sub-divided bar diagram in which the value of each component is converted into a percentage (%) of the whole (Fig. 3.5 and Table 3.6). The basic difference between these two bar diagrams is that in the sub-divided bar diagram the bars are of different heights as their total values may be different, but in percentage bar diagram bars are equal in height as each bar

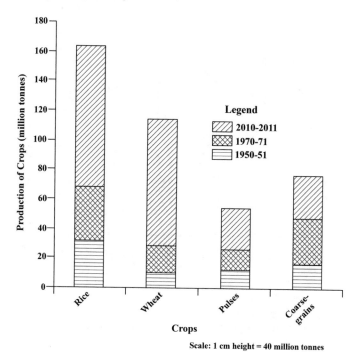

Fig. 3.4 Sub-divided bar (Production of different crops in selected years in India) *Source* Ministry of Agriculture and Economic Survey, 2010–2011 and Husain, 2014

Table 3.5 Calculations for sub-divided bar diagram (Production of different crops in India, 1950–1951 to 2010–2011)

Name of crops	Production in different time periods (million tonnes)			Scale selected	Height of the bar (cm)		
	1950–51	1970–71	2010–2011		1950–51	1970–71	2010–2011
Rice	30.8	37.6	95.32	1 cm to 40 million tonnes	0.77	0.94	2.38
Wheat	9.7	18.2	85.93		0.24	0.45	2.15
Pulses	12.5	13.4	28.0		0.31	0.33	0.7
Coarse-grains	15.5	31.4	30.0		0.39	0.78	0.75

Source Ministry of Agriculture and Economic Survey, 2010–2011 and Husain, 2014

represents 100% value (Fig. 3.5). For the data having more than one component, the percentage bar diagram will be more appropriate and convincing than the sub-divided bar diagram as the former becomes more helpful in comparison of different components.

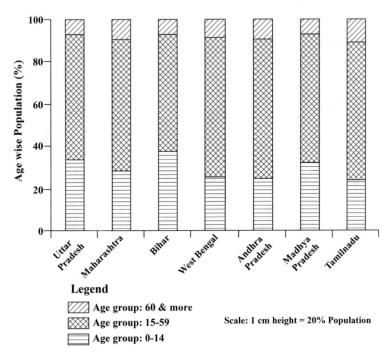

Legend
Age group: 60 & more
Age group: 15-59   Scale: 1 cm height = 20% Population
Age group: 0-14

Fig. 3.5 Percentage bar showing the proportion of population in different age groups in selected states in India *Source* Census of India, 2011

Table 3.6 Calculations for percentage bar diagram (Proportion of population in different age groups in selected states in India, 2011)

Name of the state	Age-wise (years) population (%)			Total (%)	Scale selected	Height of the bar (cm)			Total (cm)
	0–14	15–59	≥60			0–14	15–59	≥60	
Uttar Pradesh	33.7	59.5	6.8	100	1 cm to 20% population	1.68	2.97	0.35	5
Maharashtra	27.2	63.6	9.3	100		1.36	3.18	0.46	5
Bihar	37.3	55.8	7.0	100		1.86	2.79	0.35	5
West Bengal	25.5	66.3	8.2	100		1.27	3.31	0.42	5
Andhra Pradesh	24.6	66.6	8.8	100		1.23	3.33	0.44	5
Madhya Pradesh	32.1	60.8	7.1	100		1.60	3.05	0.35	5
Tamil Nadu	23.4	66.1	10.5	100		1.18	3.30	0.52	5

Source Census of India, 2011

3.4.1.2 Pyramids

It is a specific and typical type of bar diagram in which the bars are placed in such a way that it forms a pyramid-like structure. Pyramid diagrams are popularly used in different branches of geography including, population studies, urban studies, ecological or ecosystem studies etc. in different forms for the proper and accurate representation of geographical data.

Pyramids in Population Studies (Age–sex Pyramid)

In population geography, pyramid is generally used to represent the age–sex composition of the population (age–sex pyramid) of a country or region. Different age groups are shown vertically, the base representing the youngest group and the apex representing the oldest group, whereas the male and female population are shown horizontally, male population being to the left and female population being to the right of the pyramid. Generally, age groups are considered to be uniform and horizontal bars of uniform width are drawn. But, if the age groups are unequal then the width of the bars becomes unequal.

Population pyramids may be drawn in two ways: (i) based on absolute numbers of the male and female population (Fig. 3.6a) and (ii) the proportion or percentage of the male and female population with respect to the total (Fig. 3.6b). There are two possible methods for the conversion of percentage values from absolute numbers of male and female: First, the numbers of female population in each age group may be

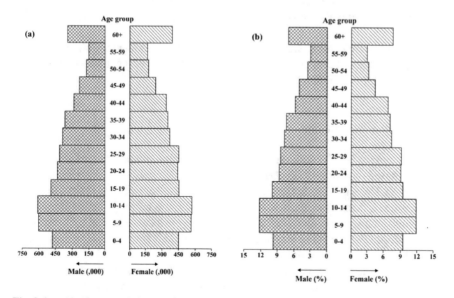

Fig. 3.6 a Absolute population pyramid and **b** percentage population pyramid

expressed as the percentage of the total female population of a country or region. The percentage of male population in each age group may be calculated in the same way (Table 3.7). Secondly, the male and female population in each age group may be expressed as the percentage of the total population of a country or region. The pyramid of absolute numbers shows the size and composition of the population of a country or region, whereas the pyramid in percentage is used to compare the age–sex composition of population of two or more countries or regions, on a single scale of which one is small and the other is big. The work participation status of the male and female population in different age groups can be easily represented by pyramid diagram.

Pyramids in Ecological Studies

The use of pyramid diagram is very popular in ecological or ecosystem studies. The structure and function of successive trophic levels, i.e. producers, primary consumers, secondary consumers and tertiary consumers may be represented graphically by means of *ecological pyramids* (Sharma 1975). In this pyramid, the producer (commonly the green plants) constitutes the base of the pyramid (Trophic level-1) and successive levels or tiers are occupied by the organisms of different consumer levels making an apex (Fig. 3.7). Based on the number of organisms, biomass and energy at different trophic levels, ecological pyramids are of three types: (i) *Pyramid of numbers*: It portrays the number of organisms at different trophic levels which commonly decreases from base to apex. (ii) *Pyramid of biomass*: It represents the total dry weight of the total amount of living organisms or matters. (iii) *Pyramid of energy*: It represents the rate of energy flow and/or productivity at different trophic levels.

All these ecological pyramids are commonly upright in shape. But, pyramids of number and biomass are sometimes inverted in shape depending upon the nature and character of the food chain of a particular ecosystem.

Pyramids in Urban Studies

In urban geography, the absolute numbers or percentage distribution of urban areas or cities based on different size classes may be represented in the form of a pyramid (Table 3.8). Here, the number or percentage of cities in each class is plotted horizontally against different size classes of the cities (plotted vertically). Thus symmetrical horizontal bars are placed on both sides of the size class column forming a pyramid-like structure (Fig. 3.8). The length of each bar is directly proportional to the number or percentage of cities it represents.

Similarly, the number or percentage of population living in different size classes of towns or cities may also be portrayed in the form of urban pyramid. Generally, urban geographers around the world, rigorously and successfully use these types of

Table 3.7 Worksheet for age-sex pyramid (Based on the population of Purba Medinipur district, West Bengal, 2011)

Age group	Absolute number ('000)		Scale selected	Length of the bar (cm)		Percentage (%)		Scale selected	Length of the bar (cm)	
	Male	Female		Male	Female	Male	Female		Male	Female
0–4	469	449	1 cm to 1,50,000 male and female	3.13	3.0	9.54	9.38	1 cm to 3% male and female	3.18	3.13
5–9	597	569		3.98	3.79	12.14	11.89		4.05	3.96
10–14	598	570		3.99	3.8	12.16	11.90		4.05	3.96
15–19	487	450		3.25	3.0	9.90	9.40		3.3	3.13
20–24	426	435		2.84	2.9	8.66	9.09		2.89	3.03
25–29	412	438		2.75	2.92	8.38	9.15		2.79	3.05
30–34	380	357		2.53	2.38	7.73	7.46		2.58	2.49
35–39	362	337		2.41	2.25	7.36	7.04		2.45	2.35
40–44	284	323		1.89	2.15	5.77	6.75		1.92	2.25
45–49	249	214		1.66	1.43	5.06	4.47		1.69	1.49
50–54	167	150		1.11	1.0	3.40	3.13		1.13	1.04
55–59	143	135		0.95	0.9	2.91	2.82		0.97	0.94
60 +	343	360		2.29	2.4	6.97	7.52		2.32	2.51
Total	4917	4787				100	100			

Source Census of India

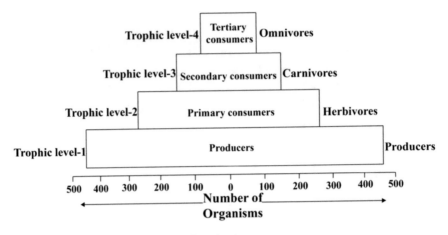

Fig. 3.7 Ecological pyramid (Pyramid of numbers)

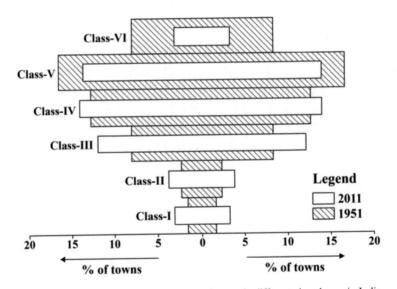

Fig. 3.8 Urban pyramid showing the percentage of towns in different size classes in India

urban pyramids to understand and explain the structural characteristics of the urban system of any country or region.

3.4.1.3 Difference Between Histogram and Bar Diagram

Though *histogram* and *bar diagram* are nearly similar in appearance, there are some specific and important differences between them.

Table 3.8 Database for urban pyramid (Size class distribution of towns in India, 2011)

Size class	Number of towns		Percentage (%) of towns		Scale selected	Length of the bar (cm)	
	1951	2011	1951	2011		1951	2011
Class-I (>100,000)	69	505	3.11	6.36	1 cm to 5% towns	0.62	1.27
Class-II (50,000–100,000)	107	605	4.82	7.63		0.96	1.53
Class-III (20,000–49,999)	363	1,905	16.36	24.01		3.27	4.80
Class-IV (10,000–19,999)	571	2,233	25.73	28.15		5.15	5.63
Class-V (5000–9999)	737	2,187	33.21	27.57		6.64	5.51
Class-VI (<5000)	372	498	16.76	6.28		3.35	1.26
Total	2,219	7,933	100	100			

Source Census of India

Histogram	Bar diagram
1. Histogram refers to the graphical representation of statistical data by rectangles or bars drawn on a horizontal baseline to show the frequency of numerical data	1. Bar diagram is the diagrammatic representation of statistical data in the form of bars to compare different categories of data
2. It indicates the distribution of different continuous variables	2. It indicates the comparison of different discontinuous or discrete variables
3. It represents quantitative data	3. It represents categorical data
4. Class boundary is shown along the x-axis and the number of observations (frequency) is shown along the y-axis	4. Time, place or other categories are shown along the x-axis while the amount or quantity of information is shown in the y-axis
5. Bars or rectangles are adjoining or continuous, i.e. there is no space between bars	5. Bars are discontinuous, i.e. there is equal space between bars
6. The height of each bar is directly proportional to the frequency of the corresponding class	6. Lengths or heights of bars are directly proportional to the amount or quantity of information they represent
7. Data are grouped together so that they turn into continuous or are considered as ranges	7. Data are taken as individual entities
8. The width of the bars is the same in equal class size but different in unequal class size frequency distribution	8. The width is same for all the bars
9. It is difficult and impossible to reorder the bars	9. Bars can be easily reordered
10. More than one frequency distribution can't be represented at a time	10. More than one component or variable can be easily represented at a time in a compound or complex bar diagram

3.4.2 Two-Dimensional Diagrams

Unlike *one-dimensional diagrams* in which only the length is considered, in *two-dimensional diagrams* the length, as well as the breadth are taken into consideration. Thus, in two-dimensional diagrams the concept of area is very significant, called area diagrams or surface diagrams. Important two-dimensional diagrams are given in the following sub-sections.

3.4.2.1 Rectangular Diagram

Rectangular diagram, an important two-dimensional method, may be used when two or more quantities are to be compared and each quantity is again sub-divided into various constituent parts (Saksena 1981) (Table 3.9). These are analogous to compound bar diagrams as the length of bars are directly proportional to the quantities they indicate but the area of the rectangles and their constituent parts are kept in proportion to the values. Generally, the rectangles are placed side by side to make them comparable. In the case of the representation of two or more sets of data, if the scale is kept the same, the computation would be easier for construction.

In rectangular diagram, data may be represented in two ways: (i) representation of the actual figures as they are given (Fig. 3.9) and (ii) by converting the actual figures into percentages (Fig. 3.10). The percentage sub-divided rectangular diagram

Fig. 3.9 Rectangular diagram showing the area of irrigated land (hectares) by different sources in India

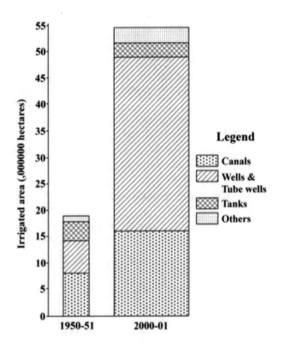

Fig. 3.10 Rectangular diagram showing the area of irrigated land (%) by different sources in India

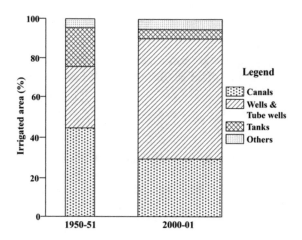

Table 3.9 Calculations for rectangular diagram (Area of irrigated land by different sources of irrigation in India)

Sources of irrigation	1950–1951			2000–2001		
	Irrigated area (thousand hectares)	%	Cumulative %	Irrigated area (thousand hectares)	%	Cumulative %
Canals	8,295	44.0	44.0	15,790	28.98	28.98
Wells and tube wells	5,980	31.7	75.7	33,275	61.07	90.05
Tanks	3,610	19.1	94.8	2,525	4.63	94.68
Others	970	5.2	100	2,900	5.32	100
Total	18,855	100		54,490	100	

Source Statistical Abstracts of India, 2005–2006 and Husain, 2014

is more popular and acceptable than the absolute sub-divided rectangular diagram as the former enables the data easily comparable on a percentage basis. Different colours or dotting or crossings may be used to distinguish the constituent parts of the rectangles.

Since the total irrigated land area in 1950–1951 and 2000–2001 are 18,855 and 54,490 thousand hectares, respectively, the width of the rectangles will be in the ratio of 18,855:54,490, i.e. 1:2.90.

3.4.2.2 Triangular Diagram

Triangular diagram is an important two-dimensional diagram which represents a series of equilateral triangles in which the size and area of each triangle ('a') is directly proportional to the quantity ('q') it indicates (Sarkar 2015). Theoretically,

$$a \alpha q \tag{3.1}$$

or a = k.q (k = proportionality constant)

If the side of the equilateral triangle having area 'a' is 'l', then

$$\frac{\sqrt{3}}{4}l^2 = a \tag{3.2}$$

[Area of an equilateral triangle with side length 'l' $= \frac{\sqrt{3}}{4}l^2$]

$$\frac{\sqrt{3}}{4}l^2 = k.q(a = k.q)$$

$$l^2 = \frac{4k.q}{\sqrt{3}}$$

$$l = \sqrt{k\frac{4q}{\sqrt{3}}} \tag{3.3}$$

So, for any item (i), corresponding to the quantity (q), the side of the equilateral triangle (l) can be represented by the following equation:

$$l_i = \sqrt{\frac{4q_i}{\sqrt{3}}} \tag{3.4}$$

For the drawing of triangles, a suitable scale should be selected carefully (Table 3.10) so that an individual triangle does not become too small or too large with respect to the given base map. Each triangle should be drawn within the boundary of the respective administrative unit of the base map, but in case of unavailability of the map, the triangles should be drawn on the same baseline maintaining uniform distance between them. The diagrammatic representation of proportional scale must contain at least three equilateral triangles showing approximately the largest, medium and smallest quantities of the given data (Fig. 3.11).

3.4.2.3   Square Diagram

Square diagram, another important two-dimensional diagram, represents a series of squares in which the size of each square is directly proportional to the quantity it signifies. Unlike rectangular diagram, in which the representation of widely varied data is difficult, in square diagram any quantity of data can be easily and simply represented.

Table 3.10 Worksheet for triangular diagram (Geographical area of selected biosphere reserves in India)

Biosphere reserve		Geographical area (sq. km)	$l_i = \sqrt{\frac{4q_i}{\sqrt{3}}}$	Scale selected	Length of the side of the triangle (cm)
Sundarban		9,630	149.13	1 cm to 80 units	1.86
Manas		2,837	80.94		1.01
Nilgiri		5,520	112.91		1.41
Gulf of Mannar		10,500	155.72		1.95
Simlipal		4,374	100.50		1.26
Panchamarhi		4,928	106.68		1.33
For proportional scale	Largest	11,000	159.38		1.99
	Medium	6,750	124.85		1.56
	Smallest	2,500	75.98		0.95

Source Geography of India by Majid Husain, 2014

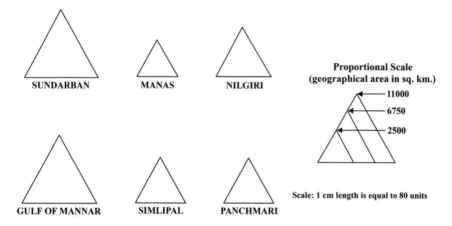

Fig. 3.11 Triangular diagram (Geographical area of selected biosphere reserves in India)

The drawing of the square diagram is based on the theory that area of each square ('a') is directly proportional to the quantity it represents ('q'). Therefore.

$$a \alpha q \qquad (3.5)$$

or, a = k.q (k = proportionality constant)

If the length of the side of a square having area 'a' be 'l', then

$$l^2 = a \qquad (3.6)$$

Table 3.11 Worksheet for square diagram (Population of selected million cities of India, 2011)

Name of the Urban Agglomeration		Population	$l_i = \sqrt{P_i}$	Scale selected	Length of the side of the square (cm)
Delhi		16,314,838	4039.16	1 cm to 1500 units	2.69
Greater Mumbai		18,414,288	4291.19		2.86
Kolkata		14,112,536	3756.66		2.50
Chennai		8,696,010	2948.90		1.96
Bangalore		8,499,399	2915.37		1.94
Hyderabad		7,749,334	2783.76		1.85
Ahmedabad		6,240,201	2498.04		1.66
For proportional scale	Largest	20,000,000	4472.13		2.98
	Medium	12,500,000	3535.53		2.36
	Smallest	5,000,000	2236.07		1.49

Source Government of India, Ministry of Information: Production Division, India (2012), New Delhi, pp. 77–78 and Geography of India by Majid Husain, 2014

[Area of a square with side length '$l = l^2$']

or $l^2 = k.q(a = k.q)$

$$l = \sqrt{k.q} \tag{3.7}$$

So, for any item ('i'), corresponding to the quantity ('q'), the length of the side of the square ('l') can be explained by the following equation:

$$l_i = \sqrt{k.q_i} \tag{3.8}$$

For the simplification of the calculation and easy understanding, $\sqrt{k.q_i}$ may be written as $\sqrt{P_i}$ in the calculation Table 3.11.

For the drawing of the squares, a suitable scale should be selected carefully so that the individual square does not become too small or too large with respect to the given base map. Each square should be drawn within the boundary of the respective administrative unit of the base map. In case of unavailability of the map, the square should be drawn on the same baseline maintaining a uniform distance between them. The diagram must contain a proportional scale having at least three squares representing roughly the largest, medium and smallest quantities of the given data (Fig. 3.12).

3.4.2.4 Circular Diagram

Like triangular and square diagram, *circular diagram* is also an important two-dimensional diagram. It consists of a series of circles in which the size or area of each circle is directly proportional to the quantity it represents. In this diagram, both

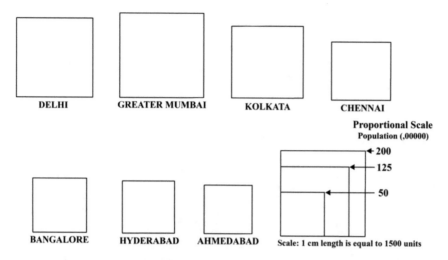

Fig. 3.12 Square diagram (Population of selected million cities of India, 2011)

the total figure and the component parts or sectors can be easily represented. The area of each circle is directly proportional to the square of its radius. The working principle for the construction of circular diagram is that the area of a circle ('a') is directly proportional to the quantity ('q') to be represented. Empirically,

$$a \, \alpha \, q \tag{3.9}$$

or a = k.q (k = proportionality constant)
    If the radius and area of a circle are 'r' and 'a', respectively, then

$$\Pi r^2 = a \tag{3.10}$$

[Area of a circle with radius '*r*' = Πr^2].
or $\Pi r^2 = k.q$ (a = k.q)

$$r^2 = k.\frac{q}{\Pi}$$

$$r = \sqrt{k\frac{q}{\Pi}} \tag{3.11}$$

So, for any item ('i'), corresponding to the quantity ('q'), the radius of the circle ('r') can be explained by the following equation:

$$r_i = \sqrt{k\frac{q_i}{\Pi}} \tag{3.12}$$

Table 3.12 Worksheet for simple circular diagram (Cropping pattern in India, 2010–2011)

Crops		Area in million hectares	$r_i = \sqrt{\frac{T_i}{\Pi}}$	Scale selected	Radius of the circle (cm)
Rice		45.0	3784.70	1 cm radius to 1500 units	2.52
Wheat		29.25	3051.32		2.03
Jowar		10.4	1819.46		1.21
Bajra		8.8	1673.66		1.12
Maize		6.4	1427.30		0.96
Gram		6.3	1416.10		0.94
Pulses		21.1	2591.59		1.73
For proportional scale	Largest	45	3784.70		2.52
	Medium	25	2820.95		1.88
	Smallest	5	1261.57		0.84

Source Government of India, Ministry of Information: Production Division, India (2012), New Delhi, pp. 77–78 and Geography of India by Majid Husain, 2014

For the simplification of the calculation and easy understanding, $\sqrt{k\frac{q_i}{\Pi}}$ may be written as $\sqrt{\frac{T_i}{\Pi}}$ in the calculation table.

Therefore, for the construction of circular diagram, radii of the circles are obtained by dividing the absolute figures (respective aggregate values) by the value of pie (Π) and taking square root (Tables 3.12 and 3.13). A suitable scale should be selected for the drawing of circular diagram so that an individual circle does not become too large or too small in size. A proportional scale must be shown diagrammatically with at least three circles roughly representing the largest, medium and smallest values of the given data (Figs. 3.13 and 3.14).

Based on the nature of data, circular diagrams are of two types:

(i) Simple circular diagram or proportional circles and (ii) sub-divided circle or compound circular diagram or angular diagram or pie diagram or wheel diagram.

Simple Circular Diagram

When the data consists of only one component (Table 3.12) then simple circles are constructed in which each circle represents a single value. The same principles are followed for the construction of *simple circular diagram* as that of constructing square diagram. The radii of the circles are taken in proportion to the square roots of the given figures following the formula mentioned earlier (Eq. 3.12). In the case of large values of the radii, they are converted to convenient small values by dividing the square roots by a suitable common value (Table 3.12). After the computation of the radii, the circles are drawn carefully keeping in mind that the centres of different

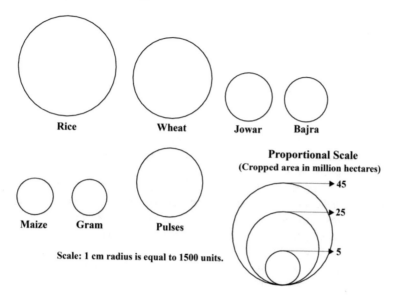

Fig. 3.13 Simple circular diagram (Cropping pattern in India, 2010–2011)

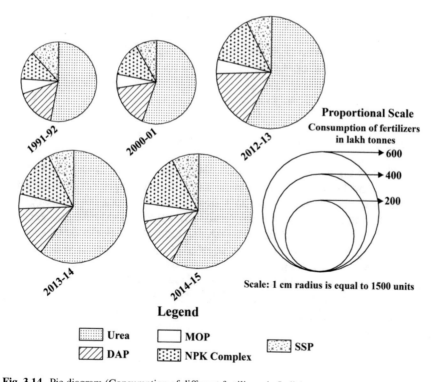

Fig. 3.14 Pie diagram (Consumption of different fertilizers in India)

circles put side by side with each other or below each other must be located on the same straight line (Fig. 3.13).

Angular Diagram or Compound Circular Diagram or Pie Diagram or Wheel Diagram

When the data is composed of a total value and two or more component parts (Table 3.13) then *compound circular diagram* or *pie diagram* is constructed (Fig. 3.14). The area of the circle represents the total value and the different sub-divisions or angular sectors of the circle represent the different component parts. In this diagram, $360°$ angles, made at the centre of the circle, correspond to the total value which is again sub-divided into a number of smaller angles or angular sectors (Fig. 3.14). The degrees of these angular sectors would be directly proportional to the values of the component parts (Table 3.13).

The angular or sectoral divisions of different component parts within the circle may be computed by the following formula:

$$s_{c_1} = \frac{360°}{q} \times c_1 \tag{3.13}$$

where s_{c_1} = degrees of an angular segment for component 1, q = total quantity of all the component parts, c_1 = quantity of component 1

$$\text{Here, } c_1 + c_2 + c_3 + \cdots + c_n = q \tag{3.14}$$

$$\text{and } s_{c_1} + s_{c_2} + s_{c_3} \cdots + s_{c_n} = 360° \tag{3.15}$$

For the drawing of the angular segments in pie diagram, it is essential to follow a number of logical principles, arrangements and patterns or sequences. As a common procedure, different angular sectors are started to be drawn from a fixed line (generally, from the radius drawn due west or north) and are arranged according to their size, with the largest at the top and the others running sequentially clockwise (Fig. 3.14). In the pie diagram, the circles and the angular segments are drawn with the help of a compass and a protector. Different angular sectors of the circles representing different components should be neatly coloured or be clearly marked by different signs and symbols in order to make the diagram attractive. A well-organized legend of colours or signs and symbols should be provided to make the diagram meaningful and understandable.

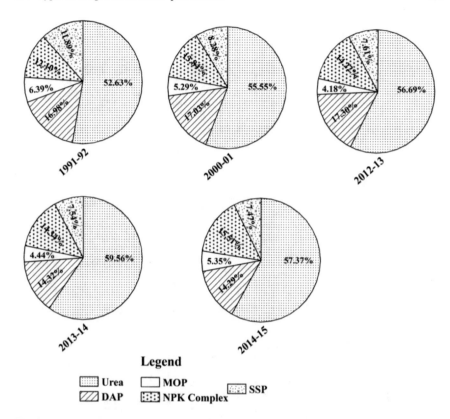

Fig. 3.15 Percentage pie diagram showing the consumption of different fertilizers in India

Pie Diagram in Percentage

In the case of comparison of data, percentage representation of *pie diagram* is more appropriate and useful than absolute representation. Because in a series of pie diagrams, it is needed to represent the larger total figure by a larger circle and the smaller total figure by a smaller circle. This type of representation involves difficulties and complications of two-dimensional comparisons. But, if the pie diagrams are constructed based on percentage value, then all the absolute totals (including larger and smaller) are considered to be 100 percentages, and hence the size of all the pie diagrams become equal (Fig. 3.15).

For the construction of percentage pie diagram, all the component values are converted into percentage with respect to the total value (Table 3.13). In such a situation, 100% value is represented by 360° angular value at the centre of the circle, and hence 1% value is represented by 3.6° ($\frac{360°}{100}$) angular value. For example, if 'P' is the percentage value of a certain component, then it will be represented by (3.6° × P) degrees as the corresponding angular value.

Table 3.13 Worksheet for pie-diagram (Consumption of fertilizers in India, lakh tonnes)

Year	Consumption of fertilizers (lakh tonnes)					Total	Radius of the circle $r_i = \sqrt{\frac{T}{\pi}}$	Scale selected	Radius of the circle (cm)
	Urea	DAP	MOP	NPK Complex	SSP				
1991–92	140.04	45.18	17.01	32.21	31.65	266.09	2910.31	1 cm radius to 1500 units	1.94
2000–01	191.86	58.84	18.29	47.80	28.60	345.39	3315.73		2.21
2012–13	300.02	91.54	22.11	75.27	40.30	529.24	4104.42		2.74
2013–14	306.00	73.57	22.80	72.64	38.79	513.8	4044.10		2.70
2014–15	306.10	76.26	28.53	82.78	39.89	533.56	4121.13		2.75
For proportional scale				Largest		600	4370.19		2.91
				Medium		400	3568.25		2.38
				Smallest		200	2523.13		1.68

Year	Consumption of fertilizers (Degree)					Total (Degree)	Consumption of fertilizers (%)					Total (%)
	Urea	DAP	MOP	NPK Complex	SSP		Urea	DAP	MOP	NPK Complex	SSP	
1991–92	189.46	61.13	23.01	43.58	42.82	360	52.63	16.98	6.39	12.10	11.89	100
2000–01	199.97	61.32	19.06	49.82	29.81	360	55.55	17.03	5.29	13.84	8.28	100
2012–13	204.08	62.27	15.04	51.20	27.41	360	56.69	17.30	4.18	14.22	7.61	100
2013–14	214.40	51.54	15.97	50.90	27.18	360	59.56	14.32	4.44	14.34	7.54	100
2014–15	206.53	51.45	19.24	55.85	26.91	360	57.37	14.29	5.35	15.51	7.47	100

Source State of Indian Agriculture 2015–16, Government of India; State Governments

Disadvantages of Pie Diagrams

Though pie diagram is frequently used as a common statistical technique, the construction of this diagram is time-consuming compared to other diagrams especially than bar diagram. Accurate reading and interpretation of a pie diagram become very difficult, particularly when the circles are divided into a large number of component sectors or the variation between these components is very little. Generally, it is not suitable to construct a pie diagram when the data is composed of more than five or six components or categories. In the case of eight or more components, it becomes very difficult and confusing to differentiate the relative quantities of them represented in the pie diagram, especially when several small sectors having approximately the same size are there. Generally, pie diagram appears upon comparison inferior to other diagrams and curves like compound bar diagram or a group of curves.

3.4.2.5 Doughnut Diagram

Like pie diagram, *doughnut diagram* displays the relationship of component parts to a whole, but it is capable of containing more than one data series (Table 3.14). In this diagram, each set of data is represented by a ring in which the first data set is displayed at the centre and the last data set towards the outside. Similar to the pie diagram, in doughnut diagram component items are represented by individual slices (Fig. 3.16). If we want to demonstrate the changes of different component parts of something, then doughnut diagram will be more appropriate than the other type of diagrams like bar or pie diagrams. Thus, it gives a birds-eye view of the relative changes in each component part of the data series.

A doughnut diagram demonstrates different category groups, series groups and series values in the form of doughnut slices. The size of each slice is directly proportional to the value it represents in proportion to the total values. In the doughnut's hole at the centre, the data labels and the totals can be displayed to make it easier to compare different segments. If the data labels are represented in percentage then each ring will total 100%.

Doughnut diagrams are of two types: simple doughnut and exploded doughnut. An exploded doughnut diagram is identical to a simple doughnut diagram but the only difference is that in the exploded doughnut, the slices are moved away from the centre of the diagram, resulting in a gap between the doughnut slices.

When the Doughnut Diagram Should Be Used

1. If we want to represent more than one data series.
2. No negative value in the data series exists.
3. When the data don't have more than seven or eight component parts.

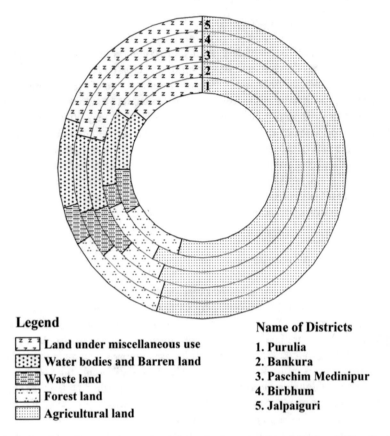

Fig. 3.16 Doughnut diagram (Area under different land uses in selected districts of West Bengal)

Table 3.14 Database for doughnut diagram (Area under different land uses in selected districts of West Bengal)

Name of the districts	Area (in thousand acres)				
	Agricultural land	Forest land	Waste land	Water bodies and barren land	Land under miscellaneous use
Purulia	336.06	75.05	49.27	68.32	87.05
Bankura	367.02	66.02	22.17	75.16	100.2
Paschim Medinipur	467	51.06	57	82	175
Birbhum	304	47	32	52	116
Jalpaiguri	396	82.16	30	69.20	141

Advantages and Disadvantages of Doughnut Diagram

The major advantages and disadvantages of using doughnut diagram include:

Advantages

1. Multiple data sets can be easily represented in a doughnut diagram.
2. Using this diagram, we can get a birds-eye view of the relative changes of different component items within the data series.
3. Comparison of different component parts using different slices becomes easy.
4. The blank space inside a doughnut diagram can be used to show the information which the diagram actually indicates.

Disadvantages

1. Due to their circular shape, doughnut diagrams are not easy to understand, especially when they represent numerous sets of data.
2. In doughnut diagram, the volume of data is not represented accurately by the proportions of outer rings and inner rings. The data points on inner rings may come into view smaller than data points on outer rings though the actual values may be larger or the same. Because of this, it is necessary to display the values or percentages of them in data labels to make them more accurate and useful.

Difference Between Pie Diagram and Doughnut Diagram

Though *pie diagram* and *doughnut diagram* both display the relationship of component parts to a whole but these two are different under the following heads:

Pie diagram	Doughnut diagram
1. Demonstrates the size differences of component parts to a whole of one data series only. Thus, it is difficult to represent multiple data sets in pie diagram	1. Size differences of component parts to a whole of multiple data sets can be easily represented in a doughnut diagram
2. Proportions of areas of the slices to one another and to the diagram as a whole are significant to compare multiple pie diagrams together	2. Focus more on understanding the length of the arcs of rings rather than comparing the proportions of areas between slices
3. The inner cut out percentage defaults to 0 for pie diagrams	3. The inner cut out percentage defaults to 50 for doughnuts
4. Less space-efficient, as no blank space exists inside a pie diagram	4. Space-efficient, as the blank space inside a doughnut diagram can be used to show information inside it
5. Unable to give a birds-eye view of the relative changes of different component parts within the data set	5. It can give a birds-eye view of the relative changes of different component parts within multiple sets of data
6. Comparison of different component parts is difficult	6. Comparison of different component parts using different slices becomes easy

3.4.3   Three-Dimensional Diagrams

Three-dimensional diagrams are those in which three things, namely length, width (breadth) and height are taken into consideration. Those diagrams are also known as volume diagrams. Some important three-dimensional diagrams are in the following sub-sections.

3.4.3.1   Cube Diagram

Cube diagram is an important three-dimensional diagram which is suitably constructed for the representation of the items having wide differences between them, say, smallest and the largest values are in the ratio of 1:1000 (Saksena 1981). In this diagram, the volumes of all cubes would be in the same proportion as the ratio of the actual data given. The construction of cube diagram is based on the theory that the volume of cube ('v') is directly proportional to the quantity ('q') it represents. Thus,

$$v \, \alpha \, q \tag{3.16}$$

or $v = k.q$ (k = proportionality constant)
If the length of side and volume of a cube are 'l' and 'v', respectively, then

$$l^3 = v \tag{3.17}$$

[Volume of a cube with side length 'l' = l^3]
  or $l^3 = k.q (v = k.q)$

$$l = \sqrt[3]{k.q} \tag{3.18}$$

So, for any item ('i'), corresponding to the quantity ('q'), the length of the side of the cube ('l') can be explained by the following equation:

$$l_i = \sqrt[3]{k.q_i} \tag{3.19}$$

For the simplification of the calculation and easy understanding, $\sqrt[3]{k.q_i}$ may be written as $\sqrt[3]{P_i}$ in the calculation table.

For constructing cube diagram, at first the cube roots of the data should be calculated with the help of logarithms. Then the logarithmic figures will be divided by the value 3 and the antilog of this value will indicate the cube root. By this technique, the sides of the cubes should be made in proportion to the cube roots of the given figures. If the sides of the cubes are large enough, then they should be reduced to a convenient size by dividing the values of cube roots by a common value.

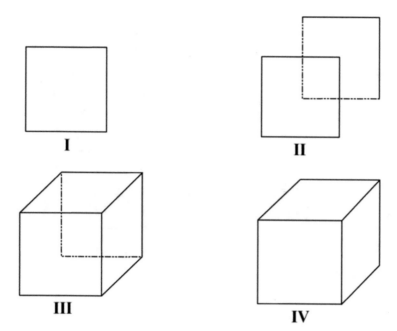

Fig. 3.17 Steps of construction of cube diagram

Steps to Construct Cube Diagram

Following steps should be followed to construct cube diagram:

1. At first, a square should be drawn with the length of the side of the cube to be portrayed (Fig. 3.17I).
2. Another square of the same size should be drawn with its lower-left corner coinciding with the centre of the first square. Thus the corresponding sides of the two squares become parallel to each other (Fig. 3.17II).
3. Then the left and right upper corners and lower right corners of both the squares should be joined by straight lines (Fig. 3.17III).
4. Lastly, the left-hand side and the lower side of the second squares should be erased and the resultant figure should be a cube (Fig. 3.17IV).

Scale for the drawing of cube diagram should be selected in such a way that none of the individual cubes is too large or too small in size. In case of unavailability of map, the cubes should be drawn on the same baseline with equal intervening space. A proportional scale must be shown diagrammatically with at least three cubes roughly representing the largest, medium and smallest values of the given data (Fig. 3.18).

Table 3.15 shows the population of the main seven tribal groups in India according to the 2011 census and data is represented using cube diagram in Fig. 3.18.

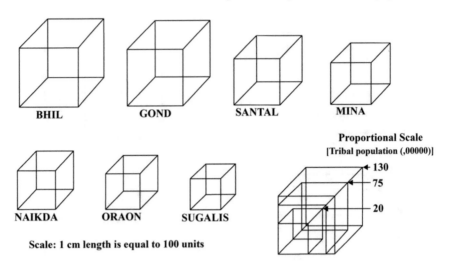

Fig. 3.18 Cube diagram (Population of main seven tribes in India)

Table 3.15 Worksheet for cube diagram (Population of main seven tribes in India, 2011)

Name of tribes		Population	$l_i = \sqrt[3]{P_i}$	Scale selected	Side of the cube (cm)
Bhil		12,689,952	233.25	1 cm to 100 units	2.33
Gond		10,859,422	221.45		2.21
Santal		5,838,016	180.06		1.80
Mina		3,800,002	156.05		1.56
Naikda		3,344,954	149.55		1.50
Oraon		3,142,145	146.47		1.46
Sugalis		2,077,947	127.61		1.28
For proportional scale	Largest	13,000,000	235.13		2.35
	Medium	7,500,000	195.74		1.96
	Smallest	2,000,000	125.99		1.26

Source Census of India

N.B. Cube roots may be calculated as follows:

$$\text{Cube root of a number} = \text{Antilog}\left(\frac{\text{Log of the number}}{3}\right)$$

3.4.3.2 Sphere Diagram

Sphere diagram is another important *three-dimensional diagram* consisting of a series of spheres which are constructed based on the principle that the volume of each sphere ('v') is directly proportional to the quantity ('q') it represents. Thus,

$$v \propto q \tag{3.20}$$

or $v = k.q$ (k = proportionality constant)

If the radius and volume of the sphere are 'r' and 'v', respectively, then

$$\frac{4}{3} \Pi r^3 = v \tag{3.21}$$

[Volume of a sphere with radius 'r' $= \frac{4}{3}\Pi r^3$]

or $\frac{4}{3}\Pi r^3 = k.q (v = k.q)$

$$r^3 = k \frac{3q}{4\Pi}$$

$$r = \sqrt[3]{k \frac{3q}{4\Pi}} \tag{3.22}$$

For any item ('i'), corresponding to the quantity ('q'), the radius of the sphere ('r') can be expressed by the following equation:

$$r_i = \sqrt[3]{k \frac{3q_i}{4\Pi}} \tag{3.23}$$

For the simplification of the calculation and easy understanding, $\sqrt[3]{k\frac{3q_i}{4\Pi}}$ may be written as $\sqrt[3]{\frac{3q_i}{4\Pi}}$ in the calculation Table 3.16.

For the drawing of sphere diagram, the scale should be selected carefully so that none of the individual spheres becomes too large or too small in size. Spheres are generally drawn within the boundary of the administrative unit of the given map. In the case of the unavailability of maps, the spheres can be drawn on the same baseline with equal distance between them. The diagram must contain a proportional scale having at least three spheres representing roughly the largest, medium and smallest quantities of the given data. Curved lines should be drawn carefully on the surface of the sphere to represent the parallels and meridians so that they appear as three-dimensional diagrams involving volumes (Fig. 3.19).

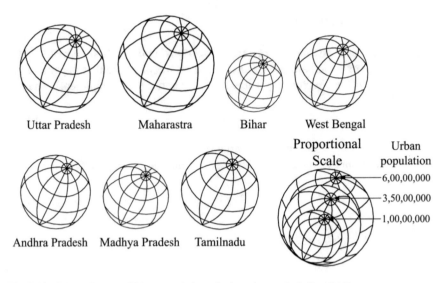

Fig. 3.19 Sphere diagram (Urban population of selected states in India, 2011)

Table 3.16 Worksheet for sphere diagram (Urban population in selected states in India, 2011)

Name of the state		Urban population	$r_i = \sqrt[3]{\frac{3q_i}{4\Pi}}$	Scale selected	Radius of the sphere (cm)
Uttar Pradesh		4,44,70,455	219.78	1 cm to 120 units	1.83
Maharashtra		5,08,27,531	229.79		1.91
Bihar		1,17,29,609	140.95		1.17
West Bengal		2,91,34,060	190.88		1.59
Andhra Pradesh		2,83,53,745	189.16		1.58
Madhya Pradesh		2,00,59,666	168.56		1.40
Tamil Nadu		3,49,49,729	202.82		1.69
For proportional scale	Largest	6,00,00,000	242.86		2.02
	Medium	3,50,00,000	202.92		1.69
	Smallest	1,00,00,000	133.65		1.11

Source Census of India

3.4.4 Other Diagrams

3.4.4.1 Pictograms

Pictograms are another very important and popular technique in which statistical or geographical data are represented by various pictorial symbols such as sacks, bales, tanks, discs etc. (Singh and Singh 1991) (Table 3.17). This is not the abstract

Table 3.17 Data for pictograms (Production of wheat in different years in India)

Year	Wheat (Ravi) production (million tonnes)	Scale selected	Number of pictorial symbols
2004–05	68.6	One pictorial symbol represents 10 million tonnes of wheat	7
2010–11	86.9		9
2011–12	94.9		10
2012–13	93.5		10
2013–14	95.9		10

Year	Number of Sacks
2004–05	
2010–11	
2011–12	
2012–13	
2013–14	

representation of data like lines or bars but it actually depicts the kind of data we want to represent. This method is more suitable and useful to the layman in representing different statistical and geographical data. In a pictogram, a number of pictures and symbols are drawn to represent different types of data.

Principles of Drawing of Pictograms

The following points should be kept in mind while a pictogram is constructed:

a. Pictorial symbols should usually be of the same size and equal in value. Each picture represents a fixed number of units or a particular quantity (Table 3.17).
b. All the pictorial symbols should be self-explanatory. For example, if we want to represent the male population then the symbol should undoubtedly indicate the male population.
c. A symbol must indicate the general idea only (like a boy, girl, truck, bus etc.) but not the individual of a species (not Hitler or Akbar etc.).
d. All the pictorial symbols drawn should be simple, clear, concise, interesting, easy to understand and easily distinguishable from every other symbol.
e. Variations in quantities or numbers should be represented by fewer or more symbols, but not by smaller or larger symbols (Table 3.17).
f. All the symbols should be drawn suitably with the size of the paper, i.e. they should not be too small or too large in size.

g.   Generally, the pictorial symbols are drawn horizontally (side by side), but they may also be drawn vertically.
h.   The quantity or the number of units represented by each pictorial symbol should be clearly mentioned.
i.   Part of a picture may be used to represent the fraction of the total value represented by each picture.

Examples To represent 60 million tonnes of wheat produced in a region, six sacks may be heaped together when one sack is supposed to represent 10 million tonnes of wheat. Similarly, to represent 80 aeroplanes in an airport, eight symbols of aeroplane may be drawn together when one aeroplane symbol is supposed to represent 10 aeroplanes.

Advantages and Disadvantages of the Use of Pictograms

The major advantages and disadvantages of the use of pictograms are:

Advantages

(1)   Pictograms are more attractive and impressive than other types of diagrams. When it is needed to attract the attention of the masses (people) such as in exhibitions, fairs etc. then pictograms are very popular in use.
(2)   Facts and events represented in a pictorial form are usually remembered longer than representation in tables or other diagrammatic forms.
(3)   Comparison of different data sets becomes easy when they are represented in pictorial form.

Disadvantages

(1)   Drawing of pictograms is very difficult as it requires some artistic sense.
(2)   Pictograms provide only the overall idea of any fact or event, but they do not offer their minute details.
(3)   In a pictogram it is required to use one symbol to correspond to a fixed quantity or fixed number of units which may also create problems. For example, if one symbol represents a five lakh population, then the question is that how many symbols are required to represent a population of 27.3 lakhs.

3.4.4.2   Kite Diagrams

It represents the change of the percentage cover of geographical phenomena or characteristics over distance. It is most frequently used to show the changes in the percentage cover of different plant species along the environmental gradient (change of environmental conditions with distance).

For example, the change of plant species from the edge of a footpath or along a sand dune transect (Fig. 3.20 and Table 3.18), along a coastline etc. can be easily represented in kite diagram.

Table 3.18 Database for kite diagram (Number of vegetation species along the sand dune transects)

Name and number of species	Distance in metre (From sea to inland)										
	0	10	20	30	40	50	60	70	80	90	100
Cough grass	63 (40%)	46 (29%)	24 (15%)	15 (10%)	8 (5%)	0	0	0	0	0	0
Dandelion	12 (6%)	11 (5%)	20 (9%)	34 (16%)	55 (26%)	32 (15%)	21 (10%)	10 (5%)	10 (5%)	5 (2%)	2 (1%)
Meadow grass	0 (0%)	0 (0%)	0 (0%)	0 (0%)	0 (0%)	12 (6%)	14 (8%)	24 (13%)	32 (17%)	45 (24%)	60 (32%)

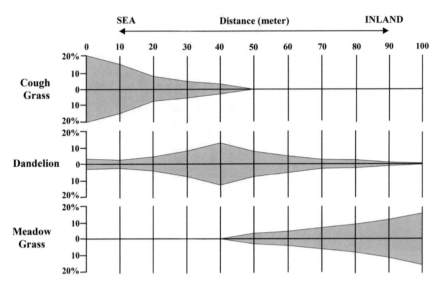

Fig. 3.20 Kite diagram showing the number of vegetation species along the sand dune transect

Procedures to Draw Kite Diagrams

Kite diagrams are drawn using the following steps:

1. At first, we need to draw a scale line to represent the distance covered in the survey.
2. One row is needed to represent each type of plant species.
3. Each and every row requires to follow the same scale and will be wide enough (sufficiently apart from the others) to allow 100% for each plant species and type.
4. Then we have to draw a line through the middle (central line) of each row representing the value '0'.
5. At each point of the survey, the percentage value is plotted on both sides above and below the central line to achieve symmetry.
6. Then the obtained points are connected for each row and it gives the diagram having a kite-like appearance.
7. The area between the kite lines are then shaded (Fig. 3.20).

Advantages and Disadvantages of Using Kite Diagrams

Use of kite diagram to represent geographical data has some advantages as well as some disadvantages:

Advantages

1. Very easy to understand and interpret.

2. Clearly shows the changes of different geographical phenomena over distance.
3. Shows the density and distribution of geographical variables.

Disadvantages

1. Not suitable for the representation of all types of data.
2. Time-consuming to plot manually.

References

Saksena RS (1981) A handbook of statistics. Indological Publishers & Booksellers
Sarkar A (2015) Practical geography: a systematic approach. Orient Blackswan Private Limited, Hyderabad, Telengana, India. ISBN: 978-81-250-5903-5
Sharma PD (1975) Ecology and environment. Rastogi Publications, Gangitri, Shivaji Road, Meerut-250002, ISBN: 978–93–5078–122–7
Singh RL, Singh RPB (1991) Elements of practical geography. Kalyani Publishers, New Delhi

Chapter 4
Mapping Techniques of Geographical Data

Abstract Map is the simplified depiction of the geographical data about the whole earth or a part of it on a piece of plane surface or paper for better understanding of their cartographic characteristics. Maps are the basic tools for geographers and researchers for the visualization of geographic data and understanding their spatial relationships. This chapter explains the basic cartographic terminologies such as Geodesy, Geoid, Spheroid, Datum, Geographic co-ordinate system, Surveying and levelling, Traversing, Bearing, Magnetic declination, Magnetic inclination etc. in a lucid manner with suitable illustrations. It includes the detailed classification and discussion of all types of maps based on their scale and purposes (contents) of preparing the map with special emphasis on Indian Topographical Sheets. All pictorial and mathematical methods of representation of relief have been explained in detail with suitable examples and illustrations. Various types of distributional thematic maps have been analyzed with suitable examples emphasizing their suitable data structure, necessary numerical calculations, methods and principles of their construction, proper illustrations and advantages and disadvantages of their use. Step-by-step and systematic discussion of the methods of construction of maps makes them easy and quickly understandable to the readers and users. Emphasis has also been given on the detailed discussion of techniques of measurement of direction, distance and area on maps.

Keywords Mapping technique · Cartographic terminologies · Representation of relief · Distributional thematic maps · Importance and uses of maps

4.1 Concept and Definition of Map

Maps are the basic tools for the visualization of *geographic data* and understanding their spatial relationships. A map is a simplified representation of the whole or part of the earth on a piece of plane surface or paper. It is a *two-dimensional depiction* of the *three-dimensional earth*. As the representation of all aspects of the earth's surface in their actual size and form is quite impossible, a map is drawn at a reduced scale. Maps are drawn in such a way that each and every point on them truly corresponds to the actual ground surface.

Map can be defined as a reduced (scaled), generalized and explained depiction (image) of objects, elements and events on the Earth or in space, constructed in a two-dimensional plane surface or paper applying mathematically defined relationships (i.e. maintaining correct relative locations, sizes and orientations). The term reduction is related to the length scale of a map, which is the ratio between the accurate length in a map and the corresponding length on ground. Generalization is an obvious outcome of the reduction, as all the particulars cannot be represented in a map in the same detail. The explanation tells us about the modes of expression and appearance through the use of legend. Map, in other words, is the representation of the earth's pattern as a whole or a part of it, or the heavens on a two-dimensional flat surface following suitable scale and projection using conventional symbols so that each and every point on it truly corresponds to the actual terrestrial or celestial position (Fig. 4.2). Three-dimensional maps can be made using the modern computer graphics only. Globes are maps portrayed on the surface of a sphere.

Map illustrates information about the earth in a simple and visual way. It serves two functions; act a spatial database and a communication device. Basic map features say to the user where an object or event is (its location) and what the object or event is (its characteristics). The amount of information to be depicted on the map depends on: (a) scale of the map, (b) projection used, (c) methods of map-making (d), conventional symbols and (e) skill and efficiency of the draughtsman or map maker etc.

4.2 Concept of Plan

A *plan* is the graphical representation of various aspects on or near the surface of the earth on a horizontal plane to a large scale. The curvature of the earth is not taken into consideration in plan. Therefore, it is suitable for smaller areas to avoid distortions related to the curvature of the earth's surface. The main purpose of a plan is to precisely and unambiguously capture all types of geometric features of an area, place, building or component (Fig. 4.1).

4.3 Difference Between Plan and Map

The differentiation between plan and map is not very easy as it is arbitrary in nature. Main areas of distinction include:

Plan	Map
1. Graphical representation of features on or near the surface of the earth on a plane or flat surface to a large scale. Scale is 1 cm = 10 m or <10 m	1. Graphical representation of the whole or part of the earth on a plane surface to a small scale compared with the plan. Scale is 1 cm = 100 m or >100 m

(continued)

(continued)

Plan	Map
2. Plans are commonly used in technical fields like architecture, engineering, planning etc.	2. Maps are commonly used to depict geography
3. Horizontal distances and directions are generally shown on a plan	3. In a number of maps, vertical distances (elevations) are also shown along with the horizontal distances and directions. For example, on a topographical map, elevations are shown by contour lines
4. A plan is drawn for small areas. For example, plan of a house, plan of a market complex, plan of a college campus etc.	4. A map is drawn for large area. For example, map of Asia, map of India, map of West Bengal etc.
5. In plan, details are given in the form of symbols	5. A map contains lots of important information of the area

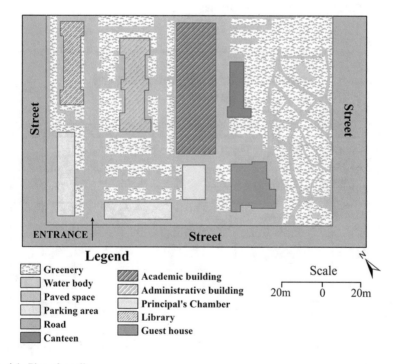

Fig. 4.1 Plan of a college campus

4.4 Elements of a Map

Several important elements are there that should be incorporated whenever a map is
prepared for the better understanding and interpretation of the map by the viewers.
A few maps may have more than this just basic information, but all maps should
contain five basic elements like *Title, Grid, Scale, Legend* and *North Arrow*. These
elements of a map have an important role to describe map details.

1.   **Title**

Title is one of the fundamental features of a map and is very important because it lets
the viewers know the general subject matter of the map and what geographic area the
map represents. A short and catchy 'title' might be appropriate if the readers have
knowledge about the theme presented on the map. The suitable title, whether small
or long, should provide an answer to the viewers to their 'What? Where? When?'
The title 'Sediment yield in global rivers' quickly says to the readers the theme and
location of the data represented in the map (Fig. 4.2).

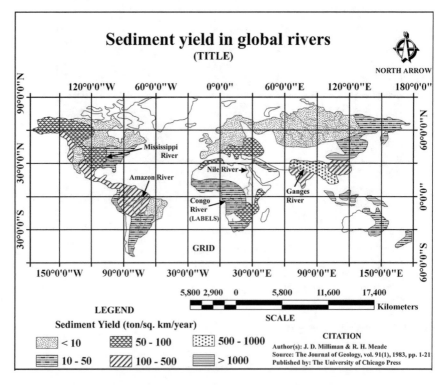

Fig. 4.2 Elements of a map (*Source* Sediment yield in global rivers, Milliman and Meade 1983)

2. **Grid**

Geographic grid system or latitude and longitude marks are really very helpful to the viewers to identify the exact location of a place or object on map. A grid is represented by a series of vertical and horizontal lines running across the map representing longitudes and latitudes, respectively (Fig. 4.2). Latitude lines (parallels) run east–west around the globe while the longitude lines (meridians) run north–south. The points of intersection of parallels and meridians are called co-ordinates. The parallels and meridians are set up with letters and numbers indicating the values of latitudes and longitudes.

On large-scale maps (objects and phenomena are shown in greater detail), the grids are generally assigned with letters and numbers. Segments (boxes) of the grid may be identified as A, B, C etc. across the top and 1, 2, 3 etc. across the left side of the map. If a stadium is located in B4 box of the grid, and it is mentioned in the index of the map, then the viewer easily finds the stadium by having a look at the box where column B and row 4 cross.

3. **Scale**

The scale represents the relation between a specific distance on the map and the actual distance in the real world, i.e. on the ground. Three main methods are there to represent map scale such as (1) Statement or Verbal Scale (i.e. 1 cm on map is equivalent to 5 km on ground or 1 inch on map is equivalent to 10 mile on ground etc.), (2) Numeric or Ratio Scale (i.e. 1:10,000, it means that each one map unit represents 10,000 units on the real world or a distance of one inch on the map equals 10,000 inches on real world or a distance of one cm on the map equals 10,000 cm on real world) and (3) Graphical Scale (ratio of map distance and ground distance can be shown graphically in the form of a scale bar like linear scale, diagonal scale etc.) (Fig. 4.2). In case of computer-generated maps, the graphical form of representation of scale is generally preferred. The maps that are drawn without following scale are required to have a 'Not to scale' notation.

4. **Legend**

Cartographers use different symbols and colours to represent various geographic features. For example, black dots to represent cities; various sorts of lines to represent national and international boundaries, roads, rivers etc.; green colour for forest; blue colour for water etc. The legend is the key element of a map describing all unknown or unique symbols and colours on the map. The legend acts as the decoder and explains what the various symbols and colours used in the map represent. Descriptions specifying any colour combinations, symbology or categorization are clearly explained in legend. Without the legend, it would be difficult for the viewers to understand the symbols and colours used in the map.

For example, in a land-use/land-cover map, various land-use and land-cover categories are represented by different colours. The map would make no sense regarding the land-use/land-cover pattern to the viewer until proper legend is given on the map. In Fig. 4.2, the legend helps the viewer to understand the amount of annual sediment yield in different river basin areas in the world.

5.   **North Arrow**

The north arrow or compass rose indicates the orientation of the map, i.e. to indicate the cardinal points (also called cardinal direction) of north, south, east and west (four main points of a compass; detail discussion is given later) and maintain a connection to the data frame (data frame is the part of the map displaying the data layers). As that data frame is rotated, the north arrow also rotates with it. It helps the viewer to recognize the right direction of the map as it is related to due north (cardinal direction may also be indicated by first putting the word "due"). Though few exceptions are there, but in most of the maps due north tends to be oriented towards the top of the sheet (Fig. 4.2).

Other important elements of a map include:

6.   **Inset map or Locator**

An Inset map or locator is a smaller map placed on the main map to further aid the viewer. It is one type of reference map, which might show the relative location of the main map. An inset map might also display a detailed, zoomed in portion of the main map.

7.   **Labels**

The words identifying the locations on the map are called labels. They show different places (streets, rivers etc.) and establishments with their distinct names (Fig. 4.2).

8.   **Citation**

The citation section of a map represents the metadata (description) of the map. This is the area that contains information such as data sources, date of creation and map projection etc. Citations facilitate the users to determine the use of the map for their own purposes (Fig. 4.2).

4.5   History of Map-Making

Maps are not the discovery of the modern human being. The history of mapping the earth is as older as the history of mankind itself. With the advancement of time, maps have appeared in different forms to the human being. Most of the earliest maps were generalized, pictorial, displaying the ideas by a rough sketch or picture, without any scale or accuracy regarding relative position or size. Over 3,000 years ago, the Egyptians made more realistic maps for the first time predominantly showing the land boundaries for the purpose of making proper assessment of revenue.

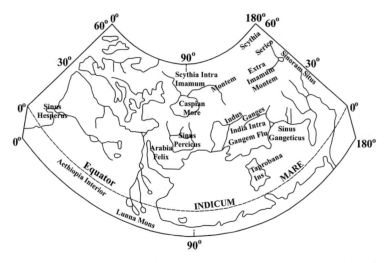

Fig. 4.3 World map of Ptolemy

4.5.1 Ancient Age

The ancient Greeks scholars and geographers are considered as the founders of modern scientific cartography as their contributions and achievements in this field were not excelled till the sixteenth century. Recognizing the shape of the earth as a spheroid with its equator, tropics and poles; identification of different climatic zones on earth; development of the grid system of latitude and longitude (projection and graticules); measurement of the circumference of the earth are all achievements of the Greek scholars and geographers like Aristotle, Ptolemy, Eratosthenes etc. Eratosthenes calculated the circumference of Earth using mathematical principles and observations of the sun. The Greek cartography culminated in the works of Claudius Ptolemy. He developed map-making techniques so precisely that would not be seen again until the fifteenth century (Fig. 4.3). The book 'Geographia' is the combined result of all his understanding and knowledge about the world. The Roman scholars and geographers, on the other hand, paid little interest in mathematical geography. They accepted the concept of old disk maps of the early map-makers, which fulfilled military and administrative purposes only.

4.5.2 Mediaeval Age

During the Dark Ages in the development of geographical knowledge in Europe, Arab scholars contributed a lot to keep the scientific cartographic techniques alive. They followed the ideas and methods of the Greek scholars and made some improvements in the map-making processes and showed, at least, the Islamic world accurately. They accepted and preserved the ideas and works of Ptolemy and interpreted them in Arabic language. Arab cartographers prepared the most reliable map of the Western world for the first time. During this period, Arab cartographers applied complex mathematical and astronomical principles and formulae for the determination of different map projections. In 1154, Al-Idrisi, a well-known scientist and cartographer prepared a map of the world, which was more accurate and acceptable than the world maps prepared by the European cartographers. The world map of Al-Idrisi consisted of the representation of the entire Eurasian continent, together with the Arabian Peninsula, the island of Sri Lanka, Scandinavia, the Black and Caspian Seas.

4.5.3 Modern Age

During fifteenth century, *cartographic knowledge* improved significantly in Europe. At the same time, sailors started travelling farther on the oceans and prepared the maps of newly discovered lands and coastlines in detail. European explorers discovered much of the Americas in sixteenth century, Australia during seventeenth century and Antarctica in the early nineteenth century. During this time, more accurate maps of the whole world were started to be accumulated. In the nineteenth century, cartographic knowledge increased dramatically due to the improvement of a printing process called lithography, which allowed the cartographers to prepare several number of maps accurately with less manpower and expense. Use of computers, photography and colour printing, all improved map-making even more. Within few decades, there was a drastic change in the relationship between people and maps. For example, inspite of using paper street maps, many people have started to navigate using GPS units, which communicate with space satellites to find out their exact position on Earth. The satellite imageries and aerial photographs have supplemented the ground method of survey, and the uses of these techniques have motivated the map-making processes in the nineteenth and twentieth centuries. Digital format of maps is capable of representing the Earth in three dimensions, eliminating the drawbacks of the earlier two-dimensional flat maps. At present, approximately, the whole earth surface has been mapped with great precession, and the information of the same is available instantly to anyone having an internet connection.

4.5.4 Contributions of Indian Scholars

The process of map-making in India was started during the Vedic period when the expressions of astronomical realities and astrophysical revelations were made. All the expressions were accumulated into 'sidhantas' or laws in standard agreements of Arya Bhatta, Bhaskara and Varahamihirá and others. During ancient period, Indian scholars divided the existing world into seven '*dwipas*' (*Jambu Dwipa, Kusa Dwipa, Plaksa Dwipa, Puskara Dwipa, Salmali Dwipa, Kraunca Dwipa and Saka*

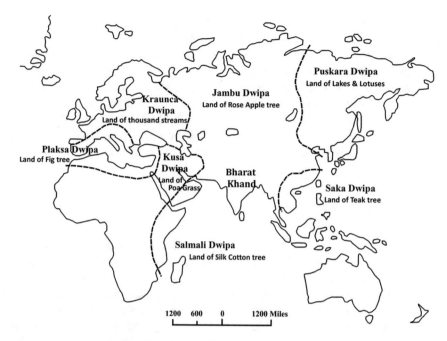

Fig. 4.4 Location and extent of Dwipic world as conceived in Ancient India [PURANAS]

Dwipa) [https://geographyandyou.com/the-geography-that-was-india/] (Fig. 4.4). In
Mahabharata, a round world surrounded by water was considered to exist.

Survey and map-making are integral parts of the revenue collection procedure in
the mediaeval period in India. The intensive topographical surveys for the preparation
of up-to-date maps of the entire country were taken up with the setting up of the
Survey of India in 1767, which culminated with the map of Hindustan in 1755.

4.6 Methods of Mapping the Earth

There are various methods by which earth surface can be mapped. These methods
include:

(1) Actual field survey using different instruments like prismatic compass, dumpy
     level, theodolite, GPS receiver, total station etc.
(2) Collecting the photographs of the surface of the earth.
(3) Freehand sketches and diagrams.
(4) Preparation of maps using modern computer techniques.

Satellite and remote sensing techniques are also widely being used for mapping
large areas of the surface of the earth more precisely.

The method of mapping the earth depends on different factors such as the size of the area to be mapped, degree of accuracy intended for, the amount of details required etc. On a topographical map, each and every point on it bears a perfect relationship with the corresponding point in the real world. Though the photographs are considered as the true representatives of the earth's pattern but they are mostly appeared in pictorial form and represent very small portion of the earth's surface at a time. Remote sensing images and satellite data are available to us to prepare the most precise and detailed maps for any region. In free-hand sketch maps and diagrams, the scale is less accurate compared with other techniques of mapping the earth surface.

4.7   Cartography

Cartography and geography are closely related with each other because both are concerned with the Earth and its life. Cartography is defined as the science and art of making maps and charts. In other words, it is the science, art and technology of making maps, charts, plans and other means of graphical representation usually on a flat surface. It may involve the superimposition of social, cultural, political or other non-geographical aspects onto the depiction of a geographical area. In broad sense, the term cartography includes all the steps like planning, field surveys, aerial photography, photogrammetry, editing, colour separation and multicolour printing required for the making of the maps and charts. Mapmakers, on the other hand, have a tendency to restrict the term to the map-finishing processes, in which the main manuscript is modified, and colour separation plates are made ready for final printing. Cartography unites the science, arts and techniques and makes it on the basis that reality can be portrayed in manners that communicate spatial data perfectly and efficiently.

The conventional analog technique of map-making has been replaced by digital technique, which is capable to produce dynamic interactive and editable maps. Cartographic techniques have undergone rapid transformations in the last two decades. Modern cartography is not only limited to the drawing of maps and charts rather it is concerned with data manipulation, data capture, processing of image and visual display of the data. Computer-based mapping systems have opened a new and exciting dimension in cartographic techniques, and existing traditional methods are required to be developed with new modern skills, efficiencies and knowledge. The fundamental character of cartography has been transformed with the advancing technologies, providing cartographers with developed modern techniques for displaying and communication of spatial information.

4.8 Key Concepts of Cartography

4.8.1 Geodesy

Over the last few decades, '*Geodesy*' has evolved from a simple surveying technique to a complex toolbox of methods, available to the scientific researchers and experts to accurately determine the positions on Earth. The word 'Geodesy' is derived from the ancient Greek word '*geodaisia*', meaning thereby 'division of Earth'. It is a branch of science (especially of applied mathematics) deals with the accurate measurement and understanding of the geometric shape of the earth, its orientation in space and gravity field. In other words, Geodesy can be defined as a branch of science that deals with the precise measurement and depiction of the Earth, considering its gravitational field, in a three-dimensional time-varying space. In *Geodesy*, the three most important properties of the Earth including its geometric shape, its orientation in space, and its gravity field along with the changes of these properties with time are accurately and scientifically measured and explained. It includes the large-scale precise measurements of various aspects of the Earth's surface and their treatment to determine the geographic positions of points or places on the Earth's surface considering the size and shape of the earth and equivalent measurements for other planets. In general, Geodesy is supposed to be related only with the measurements and representations of large areas of the Earth's surface, whereas in surveying in smaller areas, the impact of the curvature of Earth is negligible and can be ignored. Subject matter of Geodesy is strongly linked with geophysics and geodynamics. Gravity field of the Earth has an effect on geodetic measurements and determines the shape of the Earth through the Geoid.

Geodesy can broadly be divided into three main branches, although other subdivisions are considered as well:

1. Geometric Geodesy.
2. Physical Geodesy.
3. Satellite Geodesy.

4.8.1.1 Geometric Geodesy

It is purely geometrical science and concerned with describing the location of points or places in terms of geometry (shape and size) of the earth. In this science, geometrical relationships are measured using various methods such as triangulation, trilateration, electronic surveys etc. to deduce the shape and size of the earth and the precise location of particular points or places on the surface of the earth. *Geometric geodesy* uses Astro-geodetic methods and so considers the geoid. Accordingly, co-ordinate systems are important primary outcomes of geometric geodesy.

4.8.1.2   Physical Geodesy

Physical geodesy is concerned with the measurements and characteristics of the earth's gravity field and related theories in deducing the shape of the geoid and in connection with arc measurements, the size of the earth. If sufficient information regarding the earth's gravity field is available, it becomes helpful to determine and understand geoid undulations, gravimetric deflections and flattening of the earth (Burkard 1959).

4.8.1.3   Satellite Geodesy

Satellite Geodesy is the broader field of space geodesy concerned with the collection of data by means of orbiting artificial satellites for geodetic purposes. It includes the measurement of the shape and size of Earth, the location of objects or elements on its surface and the nature and form of gravity field of the Earth using artificial satellite techniques. Satellite geodesy encompasses the observational and computational methods, which help to solve and clarify various geodetic problems by accurate measurements using artificial satellites. It also includes the state-of-art techniques of locating by space methods, i.e. Global Positioning System (GPS), Very Long Baseline Interferometry (VLBI), Satellite Laser Ranging (SLR) etc.

In recent decades, applications of geodetic knowledge have rapidly increased from the measurement of plate movements and understanding earthquake hazards to incorporate research work on volcanism, landslide and weather-related hazards, climate change, water resources etc. Geodetic tools (like GPS) can also effectively be used to recognize and determine horizontal and vertical crustal movements having relation to tectonic and seismic activity.

4.8.2   Geoid

The term '*Geoid*' basically means the shape of the Earth conceptualized from its topographical characteristics. It is an idealized equipotential surface that coincides with mean sea level over the oceans in equilibrium (i.e. in constant barometric pressure at the surface; absence of air pressure, currents and tidal variations, uniform water density layers etc.) and continues in continental masses as an imaginary sea-level surface (Figs. 4.5, 4.6 and 4.7). The *geoid* is a model representing the global mean sea level and acts as a reference surface from which topographic elevations and oceanic depths are measured precisely. Therefore, geoid is a hypothetically continuous surface, which is truly perpendicular at every point or place with respect to the direction of gravity, i.e. the plumb line (Figs. 4.5 and 4.6).

We often consider the earth as a sphere, but our planet is very irregular and bumpy in reality. The larger radius of the earth at the equator than at the poles is because of long-term effects of the rotation of earth on its axis. At a smaller scale,

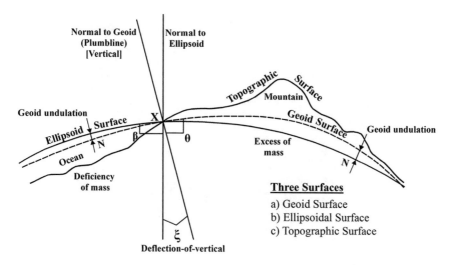

Fig. 4.5 Relation between three surfaces of the earth

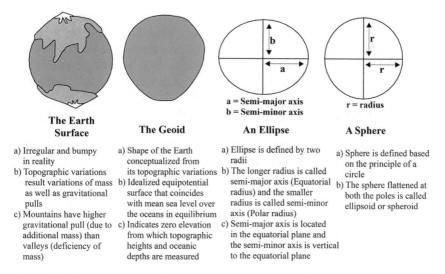

Fig. 4.6 Geoid, sphere and ellipsoid (*Source* http://physics.nmsu.edu/~jni/introgeophys/05_sea_surface_and_geoid/index.html)

topographical variation is very important in this issue, i.e. mountains have additional mass compared with valleys and as a result, the gravitational pull is comparatively stronger near mountains than valleys. All these factors, responsible for the large- and small-scale variations to the shape, size and distribution of mass of the earth, cause variations in the gravitational acceleration (i.e. the strength of gravitational pull) at different places on the earth. The shape of the earth's liquid environment (mean

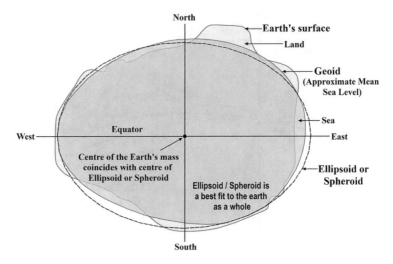

Fig. 4.7 Geoid and ellipsoid in the whole earth (Model of the earth's shape)

sea level) is determined by these variations. In the absence of the oceanic tides and currents, the mean sea level would settle onto a smoothly undulating figure, i.e. rising in high gravity areas and falling in low gravity areas. This irregular figure is considered as '*the geoid*, a level surface indicating no or zero elevation (Figs. 4.5 and 4.6). Surveyors and researchers have extended this imaginary level through the continental areas applying complex mathematical and gravity analysis on land. This model is extensively used to determine topographic height and oceanic depths with greater accuracy.

As the ideal conditions (absence of oceanic tides and currents, constant barometric pressure at the surface, uniform water density layers etc.) do not exist, locating the geoid accurately on a global scale is very difficult and enormous challenging.

4.8.3 Ellipsoid or Spheroid

The shape and size of the surface of *geographic co-ordinate system* are represented by a *sphere or spheroid*. A sphere is defined based on the principle of a circle, whereas a spheroid or ellipsoid is defined based on the principle of an ellipse (Fig. 4.6). The sphere flattened at both the poles is called as an *ellipsoid or spheroid*. In other words, ellipsoid or spheroid is a mathematical figure strongly approaching the geoid in shape and size and acts as a reference surface for geodetic surveys of a large part of the Earth (Figs. 4.5, 4.6 and 4.7) (like the Clarke spheroid of 1866 which is commonly used for geodetic surveys in the USA).

The shape of an ellipse is described by two radii. The longer radius of the ellipse is called the *semi-major axis (equatorial radius)* and is half the major axis whereas the

Fig. 4.8 Elements of an ellipse

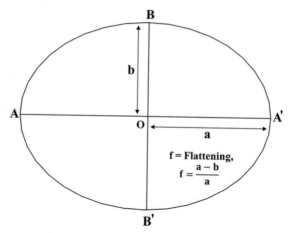

AA' = Major axis
BB' = Minor axis (Axis of Revolution)
a = Semi-major axis (One-half major axis)
b = Semi-minor axis (One-half minor axis)

shorter radius is called the *semi-minor axis (polar radius)* and is half the minor axis (Fig. 4.8). The semi-major axis is located in the equatorial plane, whereas the semi-minor axis is vertical to the equatorial plane. A spheroid is obtained by revolving an ellipse around the semi-minor axis.

A spheroid can be defined either by the semi-major axis (say '*a*') and the semi-minor axis (say '*b*') or by semi-major axis (say '*a*') and the degree of flattening (Fig. 4.8). The flattening is described as the difference between the length of two axes expressed either as a fraction or a decimal. The flattening is denoted by 'f' and is derived using the following equation:

$$f = \frac{a - b}{a} \tag{4.1}$$

The value of flattening is small, thus the measure $\frac{1}{f}$ is generally used. The spheroid parameters for the *World Geodetic System of 1984* (popularly known as WGS 1984 or WGS84):

$$a = 6378137.0 \text{ m}$$
$$b = 6356752.31424 \text{ m}$$
$$\frac{1}{f} = 298.257223563$$

The value of flattening (f) ranges between 0 and 1 where, the value '0' ($f = 0$) indicates the equal lengths of two axes, resulting in a sphere. The earth has a flattening value of around 0.003353.

The shape of a spheroid can also be defined by the square of the eccentricity of the ellipse, e^2. It is expressed by the following equation:

$$e^2 = \frac{a^2 - b^2}{a^2} \tag{4.2}$$

where, e = eccentricity of the ellipse, a = length of semi-major axis and b = length of semi-minor axis (Fig. 4.8).

The actual shape of the earth can best be represented by a spheroid, but sometimes it is considered as a sphere to make the measurements simple and mathematical calculations easier. The shape of the earth can be assumed as a sphere for those maps only having scale smaller than 1:5,000,000. In this situation, the difference in mathematical calculations between a sphere and a spheroid is negligible and is not detectable and measurable on maps. In larger scale maps with scales of 1:1,000,000 or larger, it is necessary to consider the shape of the earth as a spheroid to maintain accuracy in measurements and calculations. In between these two scales, the shape of the earth can be assumed either as a sphere or a spheroid depending on the purpose of the maps and the degree of accuracy of the data.

4.8.4 Surveying and Levelling

Cartographers consider the survey technique as the most important primary basis for the collection of data required for map-making. *Surveying* is the art and science of making such measurements as will determine the relative positions of points on, above or beneath the earth's surface by the direct or indirect measurements of direction, distance and elevation in order that the shape and extent of any portion of the earth's surface may be ascertained and delineated on a map or plan. It is basically a method of positioning points on a horizontal plane. The use of survey techniques requires skill and knowledge of mathematics, physics and astronomy.

Generally, surveyors use different instruments like compass, levelling device, theodolite, clinometer, GPS unit etc. to measure various features or topography (like position, direction, distance, slope etc.) of the land. Now, many surveyors make use of remote sensing (collection of information about the earth surface without physically touching it) technique including aerial photography (taking photographs of earth's surface from the air) to collect the data of the earth's surface. This technique has reduced the need for labours for surveyors and has allowed accurate surveying in some inaccessible (impossible to reach on foot) and restricted areas on the earth's surface.

In broad sense, levelling is a part of surveying. Levelling is the art of determining and representing the positions (heights or elevations) of different points on the surface of the earth relative to a specified datum, i.e. the method of positioning points on a vertical plane.

4.8.5 Geodetic Surveying and Plane Surveying

In general sense, based on the consideration of curvature of the earth, surveying can be divided into two main types (1) Geodetic surveying and (2) Plane surveying.

4.8.5.1 Geodetic Surveying

Geodetic Surveying (also called trigonometrical surveying) is the science of positioning and relating the position of points or objects on the earth's surface relative to each other over large distances and areas considering the curvature (the actual size, shape and gravity) of the earth. As the earth is spheroid in shape, any line joining two points on the earth's surface is actually curved or an arc of a great circle (Kanetkar and Kulkarni 1984) (Fig. 4.9a). Joining three points on the mean surface of the earth forms a spherical triangle in which sides (lines) are arcs of great circle and the angles are spherical. Thus, in Fig. 4.9b, *XYZ* is a spherical triangle and x_0, y_0, and z_0 are

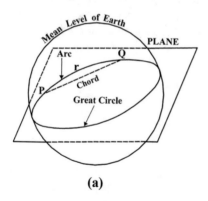

(a)

PrQ: Curved or spherical distance between PQ
(Geodetic survey)
PQ: Straight distance between PQ
(Plane survey)

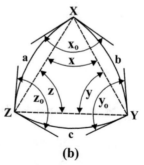

(b)

XbY, YcZ & XaZ: Sides of spherical triangle
(Geodetic survey)
XY, YZ & XZ: Sides of plane triangle
(Plane survey)

x, y & z: Plane angles
x_0, y_0 & z_0: Spherical angles

Fig. 4.9 Consideration of curvature of the earth in geodetic surveying (after Kanetkar and Kulkarni 1984)

the spherical angles. But, if the curvature of the surface of the earth is not taken into account, it becomes a plane triangle having three straight sides as represented by the dotted lines in the same figure. The resultant angles of the plane triangle are indicated by x, y and z, respectively. In geodetic surveying, the summed up value of three angles of the triangle exceeds 180°, indicating that the earth is curved. But, if the earth surface is considered to be flat, the angles would add up to 180°. Therefore, geodetic surveying is used to determine the accurate location of points on the earth's surface, of a system of widely distant points which form control stations. Engineering survey, topographical survey, cadastral survey etc. are included in geodetic surveying.

In India, *geodetic survey* is conducted by the Department of *Survey of India (SOI)*. The *Great Trigonometrical Survey (GTS)*, a pioneer project in the nineteenth century (under the leadership of George Everest and responsibility of the Survey of India) is an example of Geodetic Survey aiming to precisely and scientifically measure and mapping the entire Indian sub-continental regions. This survey covered the entire sub-continental regions with a network of triangles. Indian topographical sheets are the outcomes of this painstaking and pioneering effort.

4.8.5.2 Plane Surveying

In *plane surveying,* the surveys extend over small distance and areas, and therefore the curvature of the earth's surface is ignored. In this method, the surface of the earth is assumed as a plane; the line joining any two points as a straight line, and the angles of polygons are plane angles. The degree of accuracy of the measurement and result in this type of surveying is low compared with the geodetic surveying. American surveyors roughly put the limit at 260 sq. km. (100 sq. miles) for considering the surveys as plane surveys.

4.8.6 Datum

4.8.6.1 Vertical (Geodetic) Datum

Geodesists, surveyors and researchers use the vertical datum as the base of all kinds of geodetic surveys. They consider the vertical datum as a plane or surface of zero elevation to which altitudes can be referred to over a large geographic area. *Vertical Datum* or *datum plane* is an arbitrary level surface or line from which vertical elevations or distances are measured. It acts as a reference level or line from which level of other line or surface is calculated. This *datum* is used for the measurement of height (elevation) and depth (depression) of places above and below the mean sea level (Figs. 4.10 and 4.11). For the measurement of heights of points, *Mean Sea Level (MSL)* is considered as the most commonly accepted datum *(Orthometric datum).* MSL is the average height of seawater for all stages of tides (averaging the hourly tidal gauges) over a long period of time. In other words, a vertical (geodetic) datum

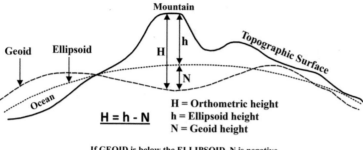

H = h - N

H = Orthometric height
h = Ellipsoid height
N = Geoid height

If GEOID is below the ELLIPSOID, N is negative
If GEOID is above the ELLIPSOID, N is positive

Fig. 4.10 Equipotential surfaces as vertical datum [Relation between orthometric height (*H*), ellipsoid or spheroid height (*h*) and geoid height (*N*)]

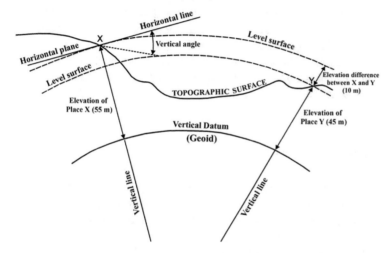

Fig. 4.11 Concept of datum

reference might consider a tidal datum as a start point. In India, the MSL at Karachi (presently in Pakistan) is considered as datum of zero elevation.

When referring to a vertical datum, we need to be aware of three main types of heights, though other types of heights may exist:

(a)   Topographic height or orthometric height (H)
(b)   Spheroid or ellipsoid height (h)
(c)   Geoid height (N).

The relationship between *Orthometric height (H), Ellipsoid or spheroid height (h) and Geoid height (N)* is shown in Fig. 4.10.

Vertical controls are established using vertical datum like Mean Sea Level (MSL) or Geoid or Earth Gravitational Model 1996 (EGM 96).

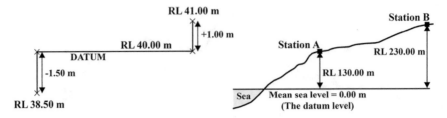

Fig. 4.12 Reduced level

4.8.6.2 Horizontal Datum

A horizontal datum is the reference value for a system of the measurement of location on earth surface. It is a model that determines the position of points on the surface of the earth in relation to the origin of latitude and longitude (Figs. 4.11 and 4.13).

For establishing horizontal control, we use horizontal datum like *World Geodetic System 1984 (WGS 84), Geodetic Reference System 1980 (GRS-80).*

When the datum is best fit to the whole earth's surface then it is called global datum but when the datum is well fit to a particular part of the earth's surface then it is called local datum.

4.8.7 Reduced Level

Reduced level is the height (elevation) or depth of a point above or below a common assumed datum or a known point (Fig. 4.12). In other words, reduced level is the vertical distance between a survey point and an assumed datum plane. It is a calculated height or level and denoted as R.L. In surveying, reduced level refers to equating heights of survey points with reference to a common datum level. It is considered as the base height and is used as the reference level to calculate the heights or depths of other important points or places.

4.8.8 Geographic Co-ordinate System

Polar angle or colatitude (φ) and *azimuthal angle (θ)* of *spherical co-ordinate system* (detail discussion in Sect. 2.2.4) do not match with the latitude and longitude of geographic co-ordinate system. The polar angle needs to be other than 0° (or 180°) to make the azimuthal value effective. For accurate matching of the two spherical angles with latitude and longitude of geographic co-ordinate system, the polar angle needs to have the value of exactly 90°. In this situation, the position vector (radius vector) points towards the positive *x*-axis in the equatorial plane and perfectly matches with 0° latitude and 0° longitude *(azimuthal = longitude)*. In *spherical co-ordinate system,* the

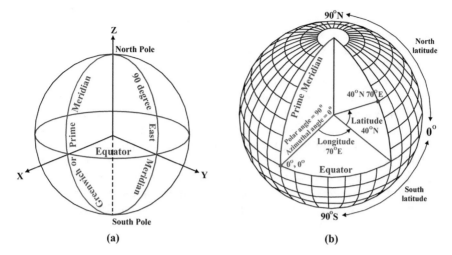

Fig. 4.13 Geographic co-ordinate system

Table 4.1 Relation between colatitude (polar angle) and latitude	Place	Colatitude/polar angle	Latitude
	North pole	0°	90°
	Equator	90°	0°
	South pole	180°	−90°

polar angle is measured from the north pole but in geographic co-ordinate, latitudinal value is measured from the equator. The relation between the *polar angle (colatitude)* and the latitude can be expressed as $\phi = 90° - \delta$, where δ is the latitude of a place and thus colatitude is considered as the complementary angle of a given latitude. Latitudes in the southern hemisphere are given a negative value and are thus assigned with a minus sign. Latitude and colatitude of a given place always sum up to 90° (Table 4.1).

Co-ordinate system on surface of the earth is known as *Geographic Co-ordinate System (GCS)*. A geographic co-ordinate system is often erroneously considered as a datum (based on a spheroid), but in true sense, a datum is only a part of it. Geographic co-ordinate system consists of a datum, a prime meridian and an angular unit of measurement. In this co-ordinate system, a three-dimensional spherical surface is used to determine the location of points on the earth surface in terms of latitude and longitude using angular unit of measurement. Latitudinal and longitudinal values are measured either in DMS (degrees, minutes and seconds) or in decimal degrees (for example, 42°21′36.36″ is equal to 42.3601 decimal degree) from the centre of the earth to a point on the earth's surface. The world is shown as a globe with latitude and longitude values (Fig. 4.13b).

In this co-ordinate system, east–west lines (horizontal lines) represent equal latitudinal values, known as parallels whereas the north–south lines (vertical lines) represent equal longitudinal values, called meridians. These parallels and meridians form a gridded network on the globe called a graticule or graticular network (Fig. 4.13b).

Equator is the line of zero latitude lying in the midway between the poles. The line representing zero longitudinal values is known as the prime meridian. In most geographic co-ordinate systems, the meridian passing through Greenwich, England is considered as the prime meridian (Fig. 4.13a, b). The point of intersection between the equator and prime meridian is defined as the origin of the graticular network (0, 0). On the basis of the compass bearings from the origin, four geographical quadrants are recognized in the entire globe. The portions of the globe above and below the equator are known as the north and south, respectively whereas the right and left sides of the prime meridian are known as east and west, respectively (Fig. 4.13b).

Latitudinal position on the earth's surface is measured with respect to the equator and the value ranges between $(-90°)$ 90°S and $(+90°)$ 90°N. Similarly, the longitudinal value is measured with respect to the prime meridian and the value ranges between $-180°$ (180°W), while moving west and $+180°$ (180°E) while moving east. The values of longitude may be equated with X and the values of latitude with Y.

4.8.9 Cardinal Points

The four main points (North, South, East and West) of a compass are known as *Cardinal directions or cardinal points*. These four directions are commonly denoted by their initials as N for North, S for South, E for East and W for West. East and west are at right angles to north and south, with east being in the clockwise bearing of turning round from north and west is directly opposite of east (Fig. 4.14).

Fig. 4.14 Cardinal points of the compass

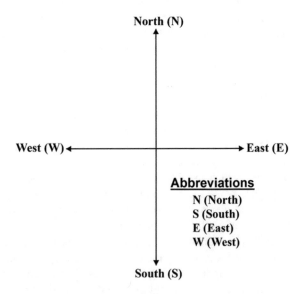

4.8.10 Map Projection

The actual shape of our planet earth is like a *three-dimensional spheroid or ellipsoid*. *Map projection* is the systematic transformation of the three-dimensional surface of the earth to a two-dimensional plane or flat surface following a suitable scale and some principles. In other words, transferring the information from the spherical surface of Earth onto a flat surface is called *map projection*. The operational process includes a dimensional transformation, i.e. a two-dimensional representation of the three-dimensional form of the earth. Due to this transformation, the shape, size and relative positions of the land masses (continents) would change. As a result, every map is characterized by some sorts of deformations, larger the area of the map, more the degree of deformation. Various aspects like the shape, size (area), distance, direction etc. can accurately be measured on earth surface, but once represented on a plane surface (projection) only a number, not all, of these features can be represented accurately. For example, a map can preserve either the accurate sizes or the shapes of the land masses, but not both.

Thus, it is not easy for the cartographers (mapmakers) to select suitable projection for the representation of the map. Two things, including the purpose of the map and the shape and size of the area of interest would be helpful to narrow down the possibilities for the selection of the appropriate projection. Depending on the purpose of the map, cartographers must decide what aspects are most important to preserve accurately and it determines which projection is suitable to use. For atlas maps and general reference maps, cartographers generally want to make a balance between shape and area distortion. But, if the map has a specific purpose, there is the need to preserve a specific spatial property (usually the shape or area or direction) for the fulfilment of the purpose.

4.8.10.1 Suitable Projections Based on Location, Shape and Purpose of the Map

Based on the location and shape of the area of interest, and purpose of the map, different types of projections are used to represent different maps. Suitable projections for the representation of various types of maps are listed in Table 4.2.

4.8.11 Bearing

In surveying, *bearing* is the technique to represent the direction of any point or place with respect to a known point or place. *Bearing* is the horizontal angle in which a line makes with a reference direction (called reference meridian), measured in a clockwise direction from the reference line. It is the number of degrees in the angle measured clockwise with respect to the reference direction. For example, the bearing of Q from P is 60° and the bearing of P from Q is 240° (Fig. 4.15).

Table 4.2 Suitable projections for different maps

Map projections	Suitable regions, shape and purposes of maps
1. Polar zenithal stereographic projection	1. Suitable for the map of the world in hemisphere because it is an orthomorphic projection (the shape of the map is truly preserved) and direction between two points is maintained properly
2. Polar zenithal gnomonic projection	2. Suitable for smaller areas around the pole. Most useful to the navigators
3. Polar zenithal orthographic projection	3. Popularly used by the astronomers for plotting different celestial bodies
4. Polar zenithal equal area projection	4. Most popularly used for the polar areas in a world atlas
5. Simple conical projection with one standard parallel	5. Suitable for smaller countries of mid-latitude or temperate regions. It is used to show the railways, roads, international boundaries etc. Trans-Siberian Railways, Canadian Pacific Railways are shown in this projection
6. Simple conical projection with two standard parallel	6. Suitable for mid-latitude countries with latitudinal extent relatively smaller than the longitudinal extent
7. Lambert conformal conic projection	7. Popularly used for aeronautical charts
8. Bonne's projection (modification of Simple conical projection with one standard parallel)	8. An equal area projection is suitable for small and compact countries with nearly equal latitudinal and longitudinal extensions, like Netherlands, France, Switzerland etc. Map of India can best be represented in Bonne's Projection
9. Polyconic projection	9. Largely used for topographical survey sheets for accurate large scale mapping
10. International map projection (modified form of Polyconic projection)	10. Used for the drawing of topographical maps of the whole world on a scale of 1:1,000,000
11. Cylindrical equal area projection	11. More suitable for equatorial land showing the distribution of tropical crops. It is also used to represent the world map
12. Mercator's projection	12. The direction is truly preserved. Thus, it is suitable for world map showing drainage pattern, wind circulation pattern, ocean circulation pattern, routes etc. where direction between places is very important. Especially used for navigation charts and maps
13. Gall's projection	13. Used for general world maps in preference to other cylindrical projections
14. Sinusoidal projection	14. Most suitable for the continents or countries in equatorial region, like Africa, South America etc. It is also used to represent the world map
15. Interrupted projection	15. By creating gaps in the world map, the map makers represent the size and shape of most of the land quite accurately
16. Mollweide's projection	16. Often used in atlases for the distribution maps of the world, as it is an equal area projection

Three types of bearings are used based on the characteristics of the reference lines or reference meridians.

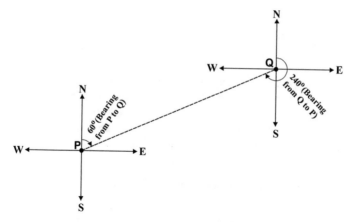

Fig. 4.15 Concept of bearing

Fig. 4.16 True and
magnetic meridian and true
and magnetic bearing

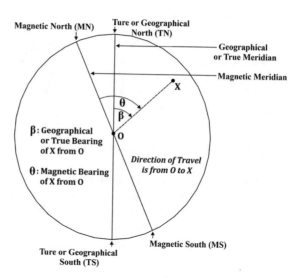

4.8.11.1   True (Geographical) Meridian and True (Geographical) Bearing

The two points of intersections of the axis of rotation of the earth and the earth's surface are known as the *true or geographical North Pole and South Pole*. The *true or geographical meridian* is a line that passes through a point, with a plane passing through the point and the true or geographical north and south poles (Fig. 4.16). In other words, it is the line that passes through the true north and the south poles. The direction of the true meridian of a point or place is constant. The true meridians through different points or places are not parallel, but they converge at the poles.

However, in case of ordinary small surveys, true meridians are supposed to be parallel to each other. The true meridian is usually employed in geodetic survey. The true meridian at any point or place can be determined either by monitoring the bearing of the sun at 12 noon or by measuring the shadow of the sun.

The true or geographical bearing of a survey line is the horizontal angle in which the line makes with the true or geographical meridian passing through one of its extremities (Figs. 4.16 and 4.19b). In other words, the horizontal angle in the clockwise direction between the true or geographical meridian and a line is known as the true bearing of the line. It is also called as an azimuth.

4.8.11.2   Magnetic Meridian and Magnetic Bearing

Magnetic meridian (also called as magnetic north and south line) is the direction indicated by a freely suspended, floating and properly balanced magnetic needle unaffected by local attractive forces (Fig. 4.16). A magnetic compass, free from other attractive forces can be used to determine the magnetic meridian. The magnetic meridian is generally used in plane survey.

The *magnetic bearing* of a survey line is the horizontal angle in clockwise direction in which the line makes with the magnetic meridian (magnetic north) passing through one of the extremities of the line (Figs. 4.16 and 4.19b). The angle made by a line with the magnetic meridian is called magnetic bearing of the line. The value of magnetic bearing ranges between 0° and 360°.

4.8.11.3   Arbitrary Meridian and Arbitrary Bearing

In certain situations (generally for small surveys), where true meridian or magnetic meridian cannot be determined, a convenient direction with respect to a well-defined permanent object or a common mark or the first line of the survey may be established and taken as a meridian during the survey. These types of reference lines are called arbitrary meridians, which becomes helpful to determine the relative positions of the survey line.

The *arbitrary bearing* is the horizontal angle in clockwise direction in which a line makes with the *arbitrary meridian* passing through one of the extremities of the line. In other words, the angle between the arbitrary meridian and a line is known as an arbitrary or assumed bearing of the line.

Based on the designations of the bearing, two types of bearing systems are used in surveying including *Whole circle bearingsystem* or *Azimuthal bearing system* (Fig. 4.17a) and *Quadrantal bearing system* or *Reduced bearing system* (Fig. 4.17b).

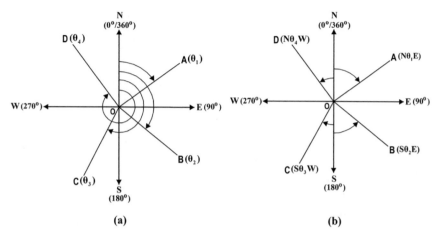

Fig. 4.17 Whole circle bearing (**a**) and quadrantal bearing (**b**)

4.8.11.4   Whole Circle Bearing System or Azimuthal Bearing System (W.C.B.)

Whole Circle Bearing (W.C.B.) refers to the bearings expressed in whole circle system. In this system, the bearing of a line is always measured in clockwise direction from the north point of the reference meridian (magnetic north) towards the line right round the circle. Thus the measured angle (bearing) is known as the whole circle bearing (W.C.B.). Therefore, the value of the bearing ranges between 0° and 360° and specified by the angle only, the noting of the letters like N, S, E or W (cardinal points) is not needed, for example, 30°, 250°, 110°30′ etc. Thus, in Fig. 4.17a, the W.C.B. of OA is θ_1, OB is θ_2, OC is θ_3 and OD is θ_4. The bearings measured with a prismatic compass or theodolites are the whole circle bearings.

4.8.11.5   Quadrantal Bearing System or Reduced Bearing System (Q.B.)

In quadrantal system, the bearing of any given line is measured either clockwise or counter-clockwise from the north point or the south point whichever is closer to the line. Hence, it is absolutely necessary to mention the point from which the angle is measured and also the direction in which it is measured. The plane around a station is divided into four quadrants and the letters N, S, E and W are used to show the quadrants. The first quadrant is denoted by the letters N. E., the second one by the letters S. E., the third one by the letters S. W., and the fourth quadrant by the letters N. W. (Fig. 4.17b). Thus, the bearing is represented by N or S first followed by angle value and E or W direction. The value of the bearing is always <90°, for example, N 30° E, S 60° E, S 45° W, N 30° W etc. Thus, in Fig. 4.17b, the Q. B. of OA is N θ_1 E, OB is S θ_2 E, OC is S θ_3 W and OD is N θ_4 W. The bearings observed with

Table 4.3 Methods of conversion of Q. B. from W. C. B

Quadrant	W.C.B. between	Rule to convert in Q.B. or R. B.
I (N. E.)	0° and 90°	N(W.C.B.)E
II (S. E.)	90° and 180°	S(180° − W.C.B.)E
III (S. W.)	180° and 270°	S(W.C.B. − 180°)W
IV (N. W.)	270° and 360°	N(360° − W.C.B.)W

a surveyor's compass are the quadrantal or reduced bearings. W.C.B. can be easily converted into Q.B. using the formulae given in Table 4.3.

For example,

1. W.C.B. $= 50°$; So, Q.B. $=$ N 50° E
2. W.C.B. $= 140°30'$; So, Q.B. $=$ S $(180° − 140°30')$ E $=$ S 39°30' E

Similarly, Q.B. can easily be converted into W.C.B.

3. Q.B. $=$ S 39°30' W; So, W.C.B. $= (180° + 39°30') =$ W.C.B. $= 219°30'$
4. Q.B. $=$ N 45°30' W; So, W.C.B. $= (360° − 45°30') =$ W.C.B. $= 314°30'$

Difference Between W. C. B. (Azimuth) and Q. B. (R. B.)

Major differences between whole circle bearing and quadrantal bearing are:

W. C. B. (Azimuth)	Q. B. (R. B.)
1. The bearings are expressed in whole circle system	1. The bearings are expressed in quadrantal system
2. Always measured in clockwise direction from the north point of the reference meridian (magnetic north)	2. Measured clockwise or anti-clockwise from the north point or the south point whichever is closer to the line, towards the east or west
3. The value of the bearing ranges between 0° and 360°	3. The value of the bearing is always <90°
4. Represented by the numerical value only, the noting of the letters like N, S, E or W is not needed	4. The bearing is represented by N or S first followed by angle value and E or W direction
5. The bearings measured with a prismatic compass or a theodolite are the whole circle bearings	5. The bearings observed with a surveyor's compass are the quadrantal or reduced bearings
6. For example, **30°, 250°, 110°30'** etc.	6. For example, **N** 30° **E, S** 60° **E, S** 45° **W, N** 30° **W** etc.

4.8.11.6   Forward Bearing and Backward Bearing

With respect to the direction of measurement, every line has two bearings. The bearing of a line measured in the direction of the progress of the survey is called

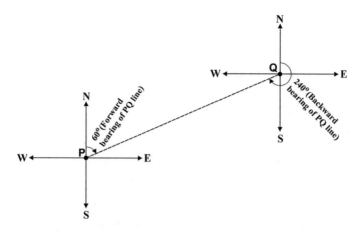

Fig. 4.18 Forward and backward bearing

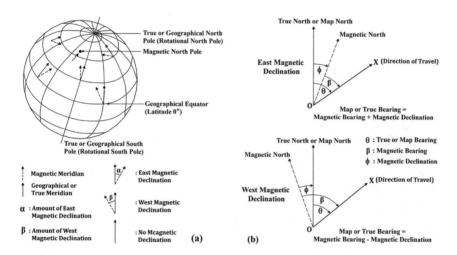

Fig. 4.19 Magnetic declination (East and west magnetic declination)

fore or forward bearing (F. B.) of the line but when the bearing is measured in the opposite direction of the progress of the survey then it is called its *back or backward bearing (B. B.)*. In line PQ (Fig. 4.18), the bearing from P to Q is the forward bearing of the line while the bearing from Q to P is called its backward bearing.

In whole circle system, the relation between the F. B. and B. B. of a line can be expressed by the rule:

$$Backward\ bearing = Forward\ bearing \pm 180°$$

(Use '+' sign, if F. B. is <180°, and '−' sign, if F. B. is >180°)
It can also be expressed in another way:

$$Backward\ bearing \sim Forward\ bearing = 180°$$

In quadrantal system, the forward and backward bearings are numerically equal but with opposite letters. Hence, the backward bearing of a line may be obtained simply by replacing N. for S. or S. for N., and E. for W. or W. for E. For example, if the forward bearing of the line AB is N 40°30′ E, the backward bearing of this line will be S 40°30′ W. Thus, the difference between the magnitude of the forward and backward bearings of a line is 0°.

4.8.12   Magnetic Declination

With few exceptions, the magnetic meridian at a point or place does not coincide with the *true or geographical meridian* at that point or place. The horizontal angle made by the *magnetic meridian* with the true meridian is called as the *magnetic declination* or magnetic variation or the declination of the needle (Fig. 4.19a, b). In other words, the angle on the horizontal plane, which is made by the magnetic meridian through a place with the true meridian through the same place, is known as the magnetic declination. When the magnetic north points to the east of the true north, the declination is said to be east or positive declination; whereas the magnetic north points to the west of the true north, the declination is said to be west or negative declination (Fig. 4.19a, b). In some places on earth surface, the needle is deflected to the east of the true north, and, in others, it is deflected west of the true north.

Since the magnetic meridian changes over time from place to place on the earth's surface, the amount and direction of the magnetic declination also vary in different localities (based on latitudes).

It is used to determine the true bearing of a line through the equation:

$$True\ bearing = Magnetic\ bearing \pm Magnetic\ declination$$

('+' when the declination is East and '−' when the declination is west.)

Lines on the surface of the earth along which the magnetic declination has the equal constant value are called *Isogonic lines* and lines along which the magnetic declination is zero are known as *Agonic lines.*

4.8.13 *Magnetic Inclination or Magnetic Dip*

Magnetic Inclination or *magnetic dip* is the angle (*I*) made by a freely suspended magnetic needle with the horizontal plane at any specific location on the earth's surface. In other words, magnetic inclination or dip (also called dip angle) is the angle in which the Earth's magnetic field lines (magnetic field vector) make with the horizontal plane of the earth's surface (Fig. 4.20). Magnetic inclination differs temporally at different places on the surface of the earth.

Between the earth's magnetic poles (magnetic North and South Pole), there is an area where the *magnetic dip or inclination (I)* is zero, called the magnetic equator. The imaginary line joining places having zero (0) magnetic dip is called magnetic equator (Fig. 4.20). There is no vertical component (*Z*) to the magnetic field at the magnetic equator. Position of magnetic equator is not fixed; rather it changes slowly from time to time. To the north of the magnetic equator, the north end of the magnetic compass needle points downward, that is the value of *I* and *Z* is positive. To the south of the magnetic equator, the south end of the compass needle points downward, indicating the negative value of both *I* and *Z*. As we move away from the magnetic equator, the value of *I* and *Z* increases. The magnetic dip is 0° at the magnetic equator and 90° at magnetic north and south poles (Fig. 4.20).

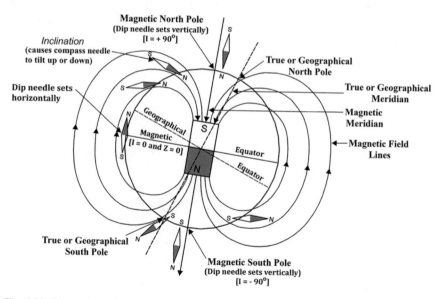

Fig. 4.20 Magnetic inclination

4.8.14 *Traversing or Traverse Survey*

Traversing or *traverse survey* is a kind of survey in which a chain of consecutive survey lines make a network and the directions and distances of the lines are measured with an angle measuring instrument and with a tape or chain, respectively (Fig. 4.21). In other words, traverse is a framework of series of connected lines whose ends have been fixed in the field, and the directions (azimuth or angle or bearing) and distances have been established from field measurements. In executing a traverse survey, the researcher or surveyor starts at a known station with a known direction (azimuth) to another point and measures directions and distances between a series of survey stations. The direction of each line of the traverse can be computed by the angular measurement and the position of each control station can be calculated by the measurements of the lengths of the lines.

A traverse may be of two types—(a) Closed traverse and (b) Open or unclosed traverse.

(a) **Open or unclosed traverse**: When the survey lines do not form a closed polygon, i.e. the lines form polygon ends elsewhere except at starting point (Fig. 4.21a), then it is known as *open or unclosed traverse.*

(b) **Closed traverse**: When the survey lines form a complete circuit, i.e. when it returns to the starting point making a closed polygon (Fig. 4.21b) or when it starts and ends at known positions or points, then the traverse is known as *closed traverse.*

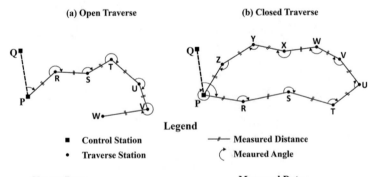

(a) Open Traverse   **(b) Closed Traverse**

Legend

■   **Control Station**   —╫— **Measured Distance**

•   **Traverse Station**   (**Meaured Angle**

Known Data: **Measured Data:**

1. Position (Latitude & Longitude) of station P   1. Length of all sides (PR, RS, ST, TU, UV, VW, WX, XY, YZ & ZP)
2. Bearing (Azimuth) of the line PQ   of the traverse
                                        2. All angles between traverse sides

Calculated Data:

1. Position (Latitude & Longitude) of all other stations (R, S, T, U, V, W, X, Y, Z)
2. Length & Bearing (Azimuth) of line between any other two stations (PS, PT, PU, PV etc.)

Fig. 4.21 Open traverse (**a**) and closed traverse (**b**)

Traversing is done by various methods based on the use of the survey instruments. The methods include:

1.   Chain Traversing
2.   Compass Traversing
3.   Plane Table Traversing
4.   Theodolite Traversing.

4.8.15   Triangulation Survey

Triangulation is the most common and important basis of trigonometrical or geodetic surveys. Triangulation is a surveying method comprising the measurement of the angles of a chain of triangles (Fig. 4.22). The main principle of triangulation method is that, by the precise measurement of the distance of one side of a triangle and the angles at both ends of the side (called baseline), the distances of other two sides and the remaining angle of the triangle can accurately be calculated. Each of the computed distances is then considered as the baseline in another triangle for the calculation of the distances to a further point, which, in turn, facilitates to make another triangle. This process is repeated as often as required to make a series of triangles connecting the point of origin to the Survey Control in the region needed (Fig. 4.22b). Generally,

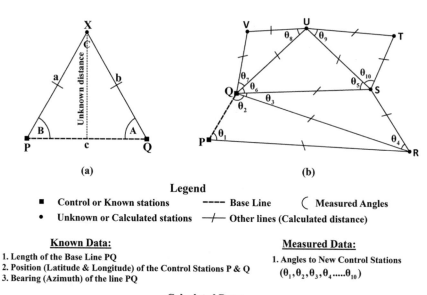

Fig. 4.22 Locating a point by angular measurement (**a**) and triangulation network (**b**)

all of the angles of every triangle are measured to minimize the errors and to furnish the data for use in calculating and verifying the accuracy of the measurements. The computed angles and distances along with the primary known position are then used for the calculation of the positions (latitude and longitude) of all other points in the triangulation network.

In Fig. 4.22a, the position (latitude and longitude) of P and Q, the distance and azimuth of the baseline PQ are known. The position of the point X with respect to P and Q can be fixed precisely by measuring the angles A ($\angle PQX$) and B ($\angle QPX$).

Therefore, $\angle PXQ = 180° - (A + B) = C$ (say).

If we consider, $PX = a$, $QX = b$ and $PQ = c$ (Fig. 4.22a), then applying the sine rule, we can measure the length of PX and QX in the following way:

$$\frac{a}{Sin A} = \frac{c}{Sin C} \tag{4.3}$$

Or,

$$a = \frac{c\, Sin A}{Sin C} \tag{4.4}$$

Or,

$$PX = \frac{PQ\, Sin A}{Sin C} \tag{4.5}$$

Similarly,

$$\frac{b}{Sin B} = \frac{c}{Sin C} \tag{4.6}$$

Or,

$$b = \frac{c\, Sin B}{Sin C} \tag{4.7}$$

Or,

$$QX = \frac{PQ\, Sin B}{Sin C} \tag{4.8}$$

4.8.16 Trilateration Survey

Trilateration is a surveying method consisting of the measurement of the distances (usually by electronic methods) of a series of triangles from a baseline, a line having known position (latitude and longitude) of two ends, distance and azimuth (Fig. 4.23). All the angles and position of the unknown point can be computed from the measured distances of the sides of the triangle. Each of the computed distances is then considered as the baseline in another triangle for the calculation of the angles to a further point, which, in turn, facilitates to make another triangle (Fig. 4.23b). This process is repeated until required numbers of triangles are obtained to cover the whole area of interest, and position of unknown points is computed in the same way. Sometimes, both distances and angles of some triangles are measured to validate the observations and improve the accuracy of the computed results. The entire network of triangles is then adjusted in order to reduce the errors due to observations.

In Fig. 4.23a, the position (latitude and longitude) of P and Q, the distance and azimuth of the baseline PQ are known. The position of the point X with respect to P and Q can be fixed precisely by measuring the distance of PX and QX.

If we consider, $PX = a$, $QX = b$ and $PQ = c$; $\angle PQX = A$, $\angle QPX = B$ and $\angle PXQ = C$ (Fig. 4.23a) then applying the cosine rule, we can measure $\angle A$, $\angle B$ and $\angle C$ in

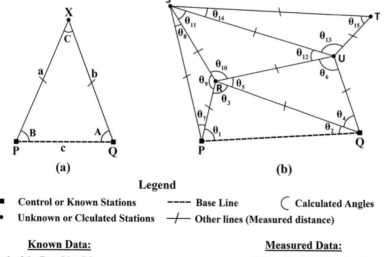

(a)   (b)

Legend

■ **Control or Known Stations**   ---- **Base Line**   ⊂ **Calculated Angles**

● **Unknown or Clculated Stations**   ─┼─ **Other lines (Measured distance)**

Known Data:	Measured Data:
1. Length of the Base Line PQ	1. Length of all Triangles sides
2. Position (Latitude & Longitude) of the Control Stations P & Q	(PR, PS, QR, QU, RU, RS, ST, TU)
3. Bearing (Azimuth) of the line PQ	

Calculated Data:

1. Position (Latitude & Longitude) of other new Stations (R, S, T, U)
2. Length & Bearing (Azimuth) between any two points (PT, PU, QS, RT etc.)

Fig. 4.23 Locating a point by linear measurement (**a**) and trilateration network (**b**)

the following way:

$$CosA = \frac{b^2 + c^2 - a^2}{2bc} \tag{4.9}$$

Or,

$$\angle A = \cos^{-1}\left(\frac{b^2 + c^2 - a^2}{2bc}\right) \tag{4.10}$$

Or,

$$\angle PQX = \cos^{-1}\left(\frac{b^2 + c^2 - a^2}{2bc}\right) \tag{4.11}$$

Similarly,

$$CosB = \frac{a^2 + c^2 - b^2}{2ac} \tag{4.12}$$

Or,

$$\angle B = \cos^{-1}\left(\frac{a^2 + c^2 - b^2}{2ac}\right) \tag{4.13}$$

Or,

$$\angle QPX = \cos^{-1}\left(\frac{a^2 + c^2 - b^2}{2ac}\right) \tag{4.14}$$

and

$$CosC = \frac{a^2 + b^2 - c^2}{2ab} \tag{4.15}$$

Or,

$$\angle C = \cos^{-1}\left(\frac{a^2 + b^2 - c^2}{2ab}\right) \tag{4.16}$$

Or,

$$\angle PXQ = \cos^{-1}\left(\frac{a^2 + b^2 - c^2}{2ab}\right) \tag{4.17}$$

4.8.17 Difference Between Triangulation and Trilateration Survey

Triangulation and *trilateration* both are important methods of survey but major differences between them include.

Triangulation	Trilateration
1. Triangulation is based on the measurement of all angles of a series of triangles	1. Trilateration is based on the measurement of all sides of a series of triangles
2. Intervisibility between stations is essential for the measurement of angles	2. The necessity of intervisibility can be avoided for the survey in small areas
3. Some check baselines are required to measure to minimize the scale errors	3. Some check base angles are measured to minimize the errors in angles (azimuth errors)
4. More internal checks are needed for a particular geometric shape	4. Internal checks are less compared with triangulation for the same geometric shape
5. Lengths of sides of the triangles are computed based on the measured angles applying sine rule	5. Angles of the triangles are computed based on the measured side lengths applying cosine rule

4.9 Types of Map

All kinds of maps prepared by the cartographers can broadly be divided into two main categories—(1) General reference maps and (2) Thematic maps.

4.9.1 General Reference Maps (General Purpose Maps)

These types of maps show various general geographic information of an area, like boundaries and names of continents, countries, towns, cities, mountains, rivers, coastlines, major transport routes etc. These are the maps that emphasize only on the location of features and all data are at same level of importance. Mainly the Government agencies such as the U.S. Geological Survey (USGS), Geological Survey of India (GSI), Survey of India (SOI) etc. prepare such type of general reference maps. These maps are very easy to understand and are mainly used for general purposes.

Example: Atlas maps, Street maps, Topographical maps etc. are the examples of General reference map. In topographical maps, the variations in surface elevation (mountains, hills, plateaux and plains), drainage, roads etc. in an area are shown in detail.

4.9.2   Thematic Maps (Special Purpose Maps)

In contrast to *general reference maps* that portray general features like coastline, terrains, drainage, roads etc., *thematic maps* emphasize on specific theme or topic. Reference maps tell us where something is whereas thematic maps focus on how something is. Thematic maps are the statistical maps focusing on the spatial distributions or patterns of one or more specific themes or spatial attributes over the surface of the earth. A thematic map is designed to display a particular theme or special topic (such as variation of temperature, distribution of rainfall, production of crops, population density etc.) connected with a specific geographic area to understand the relationship between these themes and their locations. These maps are named so because they show the attributes relating to a specific subject or theme of geography. Thematic maps may show the relationship that exists between more than one attribute or aspect. They display the attributes or statistical data in such a way that it facilitates a better understanding and explanation of the relationships between locations and the spatial patterns of the data presented. Thematic maps are called as special purpose maps, i.e. these maps are used for specific purposes.

Example: Average annual rainfall map, population density map, percentage population change map, land use/land cover map, vegetation type map, soil type map etc. are the examples of thematic maps.

Thematic maps are different from general reference maps because they do not just display different natural and man-made features like political sub-divisions, elevation, rivers, cities, highways etc. If these features are on a thematic map, they simply act as the background information and are used as reference points to improve one's understanding of the purpose and theme of the map, i.e. in thematic map, data are represented on a base map to enhance the interpretability of the theme of the map. Classification of satellite imagery (remote-sensing imagery) is an important method to produce different thematic maps. This technique is used to prepare a thematic map from the satellite imagery that the classes or features were chosen for the map are clearly identifiable in the image data. Presently, many thematic maps are easily and accurately prepared using geographic information system (GIS) technology, a computer-based system that captures, stores, manipulates, analyzes and displays all types of spatial and non-spatial data related to specific area on Earth's surface.

4.9.3   Types of Thematic Maps

Thematic maps are of different types based on how the data are visualized or represented in the map (methods of construction of the map), which, in turn, depends on the objectives of the map-making. Chorochromatic map, Choroschematic map, Choropleth map, Isarithmic map or Isopleth Map, Dot map, Flow map etc. are some examples of thematic mapping techniques that are used most often.

The above mentioned thematic maps can be categorized under two main heads.

4.9.3.1   Qualitative Thematic Map

The thematic map depicting the spatial distribution of the nominal data without any reference to the value or quantity of the element is considered as *qualitative thematic map*. Information expressed in nominal scale only reveal which class (category) a feature belongs to, i.e. the quality of the data is represented only but not the quantity.

Example: Land use/land cover map, Vegetation map, Soil map etc.

4.9.3.2   Quantitative Thematic Map

The thematic map depicting the spatial distribution of the interval and ratio data, i.e. the value or quantity of the element is referenced is considered as *quantitative thematic map*.

Example: Average annual rainfall map, population density map etc.

Detail discussion of all types of thematic maps is given in Sect. 4.12.

While classifying maps into different categories, it is essential to pay more attention to two things: (1) Scale of the map and (2) Content or purpose of the map (Table 4.4).

4.10   Types of Maps Based on Scale

Scale of a map is the key factor to control the amount of information that can be displayed in a map. The scale of a map is determined based on the extent of the area to be covered. Based on scale used, maps may be divided into two main types—(1) *Large-scale maps* and (2) *Small-scale maps*, though the third type (3) *Medium-scale maps* is also accepted.

Table 4.4 Types of maps

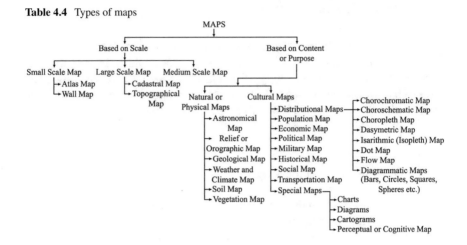

4.10.1   Large-Scale Maps

The maps that represent relative small areas of the earth's surface with greater details are called *large-scale maps*. These maps have higher representative fraction and are mainly drawn to show more details of a small area at a relatively large scale. Therefore, different features and objects in large-scale maps have appeared bigger in comparison to small-scale maps. These maps are prepared on the basis of the requirement as they are not easy to prepare because of less generalization and greater details of features.

Example: The village map, the city map, the street map of a neighbourhood etc. are the examples of large-scale maps. The topographical maps prepared at a scale of 1:250,000, 1:50,000, 1:25,000 or 1:10,000 are also the examples of large-scale maps.

Large-scale maps are again classified into two sub-types:

1.   Cadastral Map
2.   Topographical Map

4.10.1.1   Cadastral Map

The name '*Cadastral*' is derived from the French word 'Cadastre' meaning there by 'register of territorial property'. Basically, cadastral maps are drawn to register the ownership of landed properties by delineating the boundaries of fields, buildings, agricultural lands etc. They are especially prepared by the governments to realize revenue and taxes, in association with maintaining a record of ownership of landed properties.

Examples: Village maps, city plan maps etc. are the examples of cadastral maps. In our country, the village maps may be cited as example, which are drawn on a very large scale, varying from 16 inches to 1 mile (1:3,960) to 32 inches to 1 mile so as to fill in all possible details. The scale 16 inches to 1 mile is commonly used in cadastral map.

4.10.1.2 Topographical Map

Topographical maps are also prepared on a fairly large scale and are based on precise field surveys. They show general surface features in detail comprising, both natural landscapes (relief, drainage, forest and swamps etc.) and cultural landscapes (villages and towns, means of transportation and communication etc.). Uniform colours and symbols are used to represent these natural and cultural features in topographical maps. In general, the scale of topographical maps varies from 1 inch to 1 mile (1:50,000) to 1 inch to 4 mile (1:250,000). All these topographical maps have been prepared in the form of series of maps by the national mapping agencies after the precision of the methods of triangulation survey of almost all countries of the world. In India, *Survey of India (SOI)*, the *National Atlas and Thematic Mapping Organization (NATMO)* prepares the topographic maps at 1:250,000, 1:50,000 and 1:25,000 scale for the entire country. In USA, United States Geological Survey (USGS) takes the responsibility for preparing the topographical maps.

Indian Old Series Topographical Maps

1. During the time of *Great Triangulation Survey (GTS)*, the whole land area lying between 44°E, 4°N and 104°E, 40°N covering the whole of the then Indian sub-continent and a large portion of Asia was first divided into 105 uniform grids (4° × 4°) on 1 inch to 16 miles (1:1,000,000) scale and were designated by numerals 1–105, like 1, 2, 3, … 104, 105. These sheets are called million sheets or 1 M Sheets. With the present political boundary organization, there are 36 million sheets or (4° × 4°) sheets in India (Fig. 4.24a, b and Table 4.4).
2. Each million sheet is further sub-divided into 16 degree sheets (1° × 1°) on 1 inch to 4 miles scale (1:250,000) and are designated by alphabets A to P (Fig. 4.24c and Table 4.4).

3. Each degree sheet is again sub-divided in two ways. In the first way, each degree sheet is divided into four quadrant sheets (30′ × 30′) on 1 inch to 2 miles (1:100,000) scale and are designated by NE, SE, SW and NW (Fig. 4.24d and Table 4.5). In the second way, each degree sheet contains 16 inch sheets (15′ × 15′) drawn on 1 inch to 1 mile (1:50,000) scale and are designated by numerals 1–16 (Fig. 4.24e and Table 4.5).

4.   Each inch sheet (15′ × 15′) was redrawn on 1:25,000 scale using updated information obtained from both ground surveying and aerial photography and referenced in two ways. First, some of the 1:50,000 topographical maps have been printed off on old layouts, i.e. each 1:50,000 sheet contains six 1:25,000 sheets having 5′ × 7′30″ dimension and are designated by numerals 1–6 (Fig. 4.24f and Table 4.5). Secondly, each 1:50,000 sheet contains four 1:25,000 sheets having 7′30″ × 7′30″ dimension and are designated as NE, SE, SW and NW (Fig. 4.24g and Table 4.5).

Indian Open Series Topographical Maps

The Central Government of India introduced the *National Map Policy (NMP)* on 19 May 2005 and authorized the *Survey of India (SOI)* to prepare thorough guidelines for the implementation of the NMP by redesigning the existing structure of mapping layout, system of map referencing, scale and dimension of mapping and ultimately the way to access the maps. The SOI promotes an effective exchange of ideas, information and technological innovations among the data producers and users all over the globe by providing user-focused, cost-effective, consistent and quality geospatial data and information and intelligence to meet the multidisciplinary demands of national security, sustainable national development and new information markets (Sarkar 2015). The users get access to such data of the highest possible resolution at an affordable cost in the real-time environment. It will produce two different series of topographical maps, like Defence Series Maps (DSMs) and Open Series Maps (OSMs).

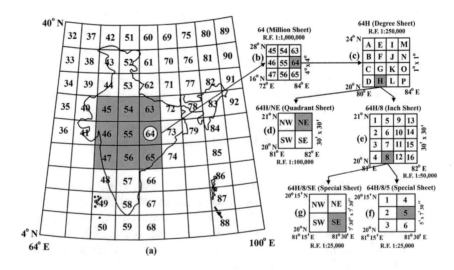

Fig. 4.24 Layout, dimension and scale of million sheets and Indian topographical maps (Old series)

Table 4.5 Layout, dimension and scale of million sheets and Indian topographical maps (Old series)

Name of sheets	Extension	Scale in FPS system	Contour interval in FPS system (ft.)	Scale in metric system	Contour interval in metric system (m.)	Reference number (examples)
Million sheet or 1M sheet	4° × 4°	1 inch to 16 miles	500	1:1,000,000	500	64
Degree sheet or quarter-inch sheet	1° × 1°	1 inch to 4 miles	250	1:250,000	100	64H
Half-inch or half-degree or quadrant sheet	30′ × 30′	1 inch to 2 miles	100	1:100,000	50	64H/NE
Inch sheet or 15′ sheet	15′ × 15′	1 inch to 1 mile	50	1:50,000	20	64H/8
Special Sheet	5′ × 7′30″ 7′30″ × 7′30″	– –	– –	1:25,000 1:25,000	10 10	64H/8/5 64H/8/SE*

* New layout

The *Defence Series Maps (DSMs)* are prepared on Everest/WGS-84 Datum and Polyconic/UTM Projection on various scales with heights, contours and full content without dilution of accuracy for the fulfilment of the requirements of defence and national security.

On the other hand, *Open Series Maps (OSMs)* are drawn on UTM Projection on WGS-84 Datum mainly to support the development mechanisms of the country. The Survey of India will ensure that civil and military vulnerable areas (VAs) and vulnerable points (VPs) are not shown on the OSMs. Thus, each of these OSMs with full topographical details will become 'unrestricted' to the users after getting one-time permission from the Ministry of Defence. All types of advanced information on SOI maps will be available at different offices, map sales centres and other *Geo-spatial Data Centres (GDC)* under SOI and also in the official website (www.survey ofindia.gov.in) of SOI. All relevant details like scale of map, date of data capturing, information content, price, mode of data dissemination etc. will be available in the aforementioned centres and website. The exact sheet number (map number) of a particular place can easily be identified by providing some basic details of the place like name of place or any other important aspect using the search engine tool available on the website of SOI.

The layout of Open Series Maps (OSMs) is as follows:

1. India is covered by 32 UTM (Universal Transverse Mercator) zones of 6° × 4° dimension drawn on 1:1,000,000 scale and are known as million sheets.

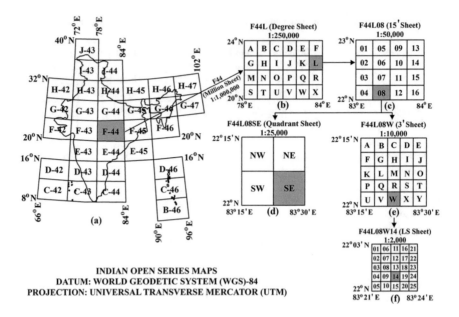

Fig. 4.25 Layout, dimension and scale of Indian topographical maps (Open series) (National Map Policy-2005, Projection-UTM, Datum-WGS-84)

Table 4.6 Layout, dimension and scale of open series maps (National Map Policy-2005, Projection-UTM, Datum-WGS84)

Name of sheets	Extension	Scale	Number of sheets	Reference number (examples)
Million sheet	6° × 4°	1:1,000,000	32	F44
Degree sheet	1° × 1°	1:250,000	24 (A–X)	F44L
15′ sheet	15′ × 15′	1:50,000	16 (01–16)	F44L08
Quadrant sheet	7′30″ × 7′30″	1:25,000	4 (NE, SE, SW, NW)	F44L08SE
3′ sheet	3′ × 3′	1:10,000	25 (A–Y)	F44L08W
LS sheet	36″ × 36″	1:2,000	25 (01–25)	F44L08W14

The sheets include B46, C42–C44, C45, D42–D44, D46, E43–E45, F42–F46, G42–G47, H42–H47, I43, I44 and J43 (Fig. 4.25a and Table 4.6).

2.   Each 6° × 4° rectangle (million sheet) is further sub-divided into 24 degree sheets of (1° × 1°) dimension following 1:250,000 scale and are designated by alphabets, A to X increasing first towards east and then towards south (Fig. 4.25b and Table 4.6).

3.   Each degree sheet is further sub-divided into 16 sheets of 15′ × 15′ dimension on 1:50,000 scale and are designated by numerals 1–16 increasing first towards south and then towards east (Fig. 4.25c and Table 4.6).

4. Each $15' \times 15'$ sheet is then divided in two ways. First, each sheet is divided into four parts (quadrant sheets) of $7'30'' \times 7'30''$ dimension following 1:25,000 scale and are designated as NE, SE, SW and NW (Fig. 4.25d and Table 4.6). Second, each sheet is divided into 25 sheets of $3' \times 3'$ dimension following 1:10,000 scale and are designated by alphabets A to Y increasing first towards east and then towards south (Fig. 4.25e and Table 4.6).

5. Each $3' \times 3'$ sheet is again divided into 25 sheets of $36'' \times 36''$ dimension on 1:2,000 scale and are designated by the numerals 1–25 increasing first towards south and then towards east (Fig. 4.25f and Table 4.6). These topographical maps are the largest scale maps published by the SOI and are called LS sheets.

4.10.2 Small-Scale Maps

The maps that are prepared to show large areas of the earth's surface with more generalization of details are called *small-scale maps*. These are called small-scale maps because they have relatively smaller scaling factor and different features and objects represented on the maps are relatively small. The larger areas, like the whole world, a continent, a country etc. are shown in a single map document with lesser details, i.e. these maps cover larger areas of the earth's surface but represent lesser amount of details of different features and objects.

Small-scale maps are of two main categories:

1. Wall Map
2. Chorographical or Atlas Map.

4.10.2.1 Wall Map

Wall maps are the maps that are generally prepared on small scale on large size paper or on plastic base to be used in class rooms or lecture halls. The scale of wall maps is smaller than that of the scale of topographical maps but larger than that of *atlas maps*. The world as a whole or in hemisphere is distinctively drawn on the wall maps. These maps may also be prepared for a continent or a country, small or large, based on the purpose and need.

4.10.2.2 Chorographical or Atlas Map

An *Atlas* can be defined as a collection of maps or a bunch of maps of the whole Earth or a region of Earth. Traditionally, most of the Atlases are found in book format, but presently, many Atlases are also found in multimedia formats.

The *Atlas maps* are prepared on a very small scale and essentially represent highly generalized picture concerning the physical, climatic and socio-economic conditions of fairly large areas on the surface of the earth. Atlas map is considered as a graphic encyclopaedia of various geographical information about the whole world, continents, countries or regions. Most of the Atlas maps show only main ranges of hills with important peaks, important rivers, main towns and cities, main transport lines (like railway tracks, national and international highways) etc. due to the limitation of space.

4.10.3   Medium-Scale Maps

All the maps with scale larger than the small-scale maps but smaller than the large-scale maps are called *medium-scale maps*. Based on the map projection systems and the plane of co-ordinates, a medium-scale map is one having a scale of 1:600,000 to 1:75,000. In medium-scale maps, earth surface features and objects are represented with greater details than the small-scale maps but lesser details than the large-scale maps.

Again, on the basis of perspective, maps are divided into the following categories based on scale:

(a)   **Geographic perspective:**

(1)   Large-scale maps: map scale greater than 1:2,00,000
(2)   Medium-scale maps: map scale 1:2,00,000 to 1:1,000,000
(3)   Small-scale maps: map scale less than 1:1,000,000.

(b)   **Geodetic perspective:**

(1)   Large-scale maps: map scale greater than 1:5,000 (possibly 1:10,000)
(2)   Medium-scale maps: map scale 1:5,000 to 1:2,00,000
(3)   Small-scale maps: map scale less than 1:2,00,000.

4.11   Based on the Purpose or Content or Function of the Map

Another important basis for the classification of maps is the purpose (why the map is prepared) or content (what types of information are represented) of the map. For example, in a political map, administrative divisions of a continent or a country are shown in detail, a vegetation map shows the spatial distribution of different types of vegetations in a region etc. In broad sense, maps are classified into two main types based on their functions or contents:

1.   Physical or natural maps
2.   Cultural maps.

4.11.1   Physical or Natural Maps

Physical or natural maps are those that represent different natural features like geology, relief, elements of weather and climate, heavenly features, drainage, soil, vegetation etc. Based on the features that are shown in physical maps, again they are of different types:

4.11.1.1   Astronomical Map

A map showing heavenly bodies or heavenly features is called *astronomical map.*

4.11.1.2   Relief or Orographic Map

A map showing the physical configuration (bulges and depressions) of the surface is called *relief map*. General topographic features of an area like mountains and valleys, plateaux, plains and drainage etc. are depicted in relief map in terms of their slope and elevation.

4.11.1.3   Techniques of Representation of Relief

Cartographers use different techniques to portray the relief of the landscape on maps, which may be grouped under three main heads:

Pictorial Methods (Portraying Visual Pictures of the Terrain)

(1)   Hachures
(2)   Hill shading.

Mathematical Methods (Indicating Heights of the Places)

(1)   Contours
(2)   Spot height
(3)   Benchmark
(4)   Trigonometrical station
(5)   Form lines.

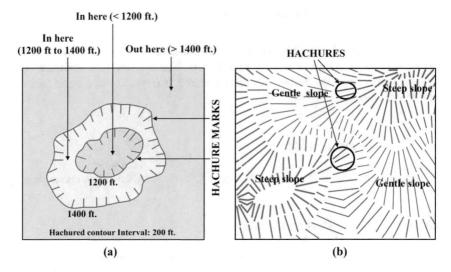

Fig. 4.26 Hachure lines or contours

Combination of Both (Combination of Visual Pictures and Actual Heights)

(1)   Contours and Hachures
(2)   Contours, Spot heights and Hachures
(3)   Contours, Spot heights and Form lines
(4)   Contours and Hill shading etc.

Hachures

Hachuring is the earliest method to represent the relief of the land on map. It represents the terrain of an area by using hachures. *Hachures* are sets of finely drawn broken (disconnected) lines indicating the direction of water flow from highland to lowland (Fig. 4.26) (Singh and Singh 1991). The direction of the *hachured lines* (also called *hachured contours*) always indicates the direction of the slope of the land. Hachures do not indicate the absolute height of places on the ground surface rather it gives an idea about the overall configuration of the ground. Therefore, hachuring is a technique to have some idea about the shape and relative steepness of slopes of the land surface.

The hachures are thicker and closely spaced on steep slopes whereas they are thin and widely spaced on gentle slopes (Fig. 4.26). Various flat areas such as mountain tops, lowland areas etc. are left blank white. For an accurate representation, the thickness of the hachured lines can be calculated mathematically, but in drawings, they are generally diagrammatic in nature. Hachures are usually applied to represent the relief in large-scale maps, these are not suitable to be used in small-scale maps. In modern cartography, the use of hachures has been reduced after being replaced by contours. Sometimes, contours and hachures are used together to make extra emphasis on the relief of the map of an area.

Demerits of Hachuring

(a)   Hachures do not indicate the actual (absolute) height of places on the surface, therefore regions, say two or more plateaux or hills can have almost similar patterns of hachuring and yet their actual heights may differ significantly.

(b)   On the map of mountainous or hilly region, too many hachures obscure other topographic details.

(c)   Drawing of hachures is time taking, exhaustive and costly as they are hand-made.

Relief Shading or Hill Shading

Hill shading is a method for creating relief maps, depicting the shape of the topography of hills and mountains with the help of shading (grey levels) on a map. Light and dark areas or shading are added to a map to illustrate the three-dimensional appearance of terrains like hills or mountains (Fig. 4.27). Cartographers generally follow the principle of 'top-left-lightning' to apply this technique to show relief on map. The relief would appear as if there is a source of light illuminated at the top-left corner of the map, creating a light-shadow effect in the map. In this method, light shading is used to highlight the areas where sunlight would strike (illuminated parts) and dark shading is used to highlight the areas where shadows would appear (obscure parts) due to the existence of hills and mountains. On the obscure parts (shadowed parts), the degree of darkness of the slope depends on the degree of its steepness, i.e. more steepness of the slope the darker is the shading, whereas, on the illuminated parts, the slope is represented by progressive light shading. Lighter shading is used to represent relatively flat areas like hilltops, plateaux, summits of ridges, valley bottoms etc. (Fig. 4.27). The variation in shadowing gives a three-dimensional effect

(a)   (b)

Fig. 4.27 Relief shading or hill shading map

to the map, which becomes helpful to understand how flat or hilly the area mapped is. The major disadvantage of using this technique is that it does not indicate the absolute height of the terrain features.

Contours

Contouring (determination of heights of different points on the ground surface and representing them in the contour map) is the standard and widely accepted method of representing relief on maps. *Contours* are imaginary lines (drawn on map) of constant elevation above the mean sea level. Contours are drawn by joining nearby places having the same height above the sea level (Fig. 4.28). Since heights are determined by actual field survey, the whole process of contouring is time consuming and costly also. But contours are easy to understand on map by the users and they do not hamper other information of the map. The general topographical characteristics of the ground surface can easily be visualized and understood with the help of contours. The difference between the values of two consecutive contour lines is called as the contour interval or the vertical interval. Brown colour is commonly used to show the contours on the topographical map.

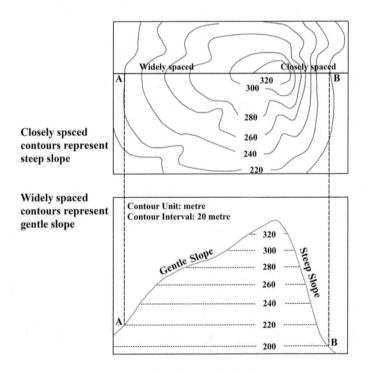

Fig. 4.28 Relation between contour spacing and steepness of slope

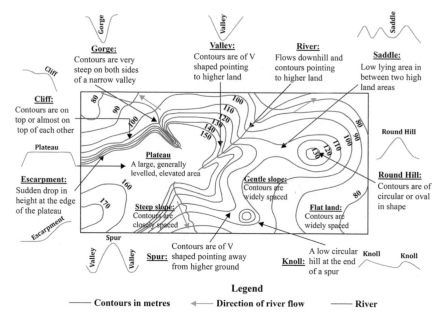

Fig. 4.29 Contour patterns of different relief features

The spacing of contours is very important to indicate the slope of the land. Closed spacing of contours indicates the steep slope of the ground whereas the wide spacing of contours (contours are farther apart) represents the gentle slope of the land (Fig. 4.28).

The pattern of the contour spacing gives us an idea about the shape or form of various topographic features on the earth's surface. Different topographic forms are represented by different patterns of contours, which are shown in Figs. 4.29 and 4.30 and described in Table 4.7.

Spot Heights or Spot Elevations

Spot height (also called *spot elevation*) is the exact height of a geographic point or place above a given datum (generally the mean sea level) on the map, fixed by actual survey. *Spot height* does not give any idea about the relief of the land. It is shown on map by black dots (•) with adjacent numerals indicating the height of point or place above mean sea-level in feet or metres, such as 1,040 (Fig. 4.31).

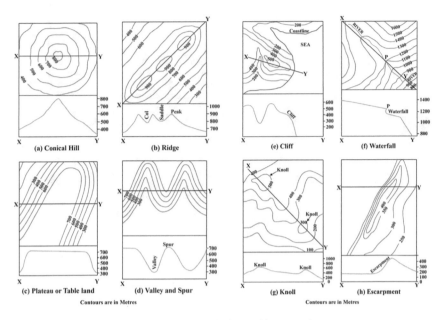

Fig. 4.30 Relation between contour pattern and topographic expression

Table 4.7 Contour patterns of typical topographic features

Topographic expression	Description	Contour pattern
Hill	An elevation having height less than 3,000 ft. (1,000 m.) above the surrounding region. Spacing of contours indicates the nature of hill (whether it is conical, flat-topped, rounded etc.). Conical hill almost rises uniformly from the surrounding land	Concentric contours having height less than 3,000 ft. (1,000 m.) spaced almost regularly. In conical hill, the circular contours become smaller to smaller towards the centre. In round top hill, the contours don't taper to a point. The central contour is comparatively large (Figs. 4.29 and 4.30a)
Ridge (col, saddle and pass or gap)	High, elongated and steep-sloped hill (or mountain) or a chain of hills with two or more peaks. A narrow and steep-sided depression between two peaks is known as col but when the depression is broader, higher and gently sloping then it is called saddle. A pass is a col (at lower level), which is used or likely to be used as a passage or route way (for footpath, road, railway etc.). All passes are cols but all cols are not passed	Contours are more or less elliptical in shape (Figs. 4.29 and 4.30b)

(continued)

Table 4.7 (continued)

Topographic expression	Description	Contour pattern
Plateau (table land)	A high land rising above the adjoining plain with relatively level surface at the top which falls down rapidly at least on one side. The plateau surrounded by mountains is called inter-montane plateau and the plateau developed at the foot hill regions is known as piedmont plateau	Innermost contours are roughly rectangular in shape (wide spacing on the plateau surface) and outer or marginal contours are closer together (Figs. 4.29 and 4.30c)
Spur and valley	A spur is a tongue of land projecting from higher ground into the lower. The low lying area of land contained between two slopes converging at their bases is called valley	Spur is represented by 'V' shaped contours in which the arms of 'V' point to higher grounds and the apex of 'V' to the lower one. Valley is also represented by V-shaped contours but in the reverse manner, i.e. the apex of 'V' pointing towards higher grounds (Figs. 4.29 and 4.30d)
Cliff	Cliff is a very steep or almost vertical exposure of landform having considerable height and overlooks a lake, a river, a sea or a plain. Cliff is generally found near sea, called sea cliff	The contours run very close to one another and ultimately merging into one on the face of the cliff (Figs. 4.29 and 4.30e)
Waterfall	A sudden and almost vertical falling of water from a considerable height in the river bed is called a waterfall. A waterfall appears when there is a sudden break of sloe, i.e. sudden difference in height of the river valley	When contours cross a stream and are very close to each another or sometimes when they even touch each other (Figs. 4.29 and 4.30f)
Knoll	A very low isolated hill that is commonly round in shape is called Knoll. On plain regions, knolls are often formed by mound of gravels and afford sites for settlements	Ring or concentric pattern of contour lines and the presence of lower ground all around (Figs. 4.29 and 4.30g)
Escarpment (scarp)	Escarpment is the abrupt (sudden change of height), comparatively long and regular, steep face of any hill ridge. Escarpment may be defined as a ridge having a steep scarp on one side and a gentle slope on the other. It is formed either as a result of faulting or erosion	Contours are closely packed (Figs. 4.29 and 4.30h)

(continued)

Table 4.7 (continued)

Topographic expression	Description	Contour pattern
Gorge	Gorge (derived from the French word gorge, meaning thereby neck or throat) is a very steep and narrow valley at higher altitudes developed by river erosion. Although gorge and canyon are used to describe deep and narrow valleys with a river or stream running along their bottom, but a canyon is often larger than a gorge	The contours converge closely in the river course, i.e. the contours become closer to each other on both sides of the stream (Fig. 4.29)
V-shaped valley	These valleys are commonly developed by the erosional activities of the running surface water or stream. From the valley line, the ground gradually slopes up	The contours that cut the valley line are closer together near the head of the valley than further down. The 'V' contours are sharper in the upper region and they widen out gradually until the V shape of the contours disappears at the mouth. The spacing of contours indicates the concave slope (angle of slope decreases downward) (Fig. 4.29)
U-shaped valley	These valleys are commonly formed by the erosional activities of the glacier. The valley has a flat floor and steep parallel sides	Contours are close and parallel running on two sides of the flowing stream or river (glacial valley is later occupied by the river), leaving a broad flat valley floor. The spacing of contours indicates the convex nature of the slope (angle of slope increases downward) (Fig. 4.29)

Benchmark and Its Types

Benchmark is a fixed reference point whose height with respect to a common datum is well known to us. These are marks placed on relatively permanent material object (natural or artificial) like, rocks, buildings etc., indicating the height above mean sea level (datum), established during the time of actual survey. On maps, these are specified by the letters B.M. followed by the number representing the height of the mark in metres or feet (such as BM120) above sea level (Fig. 4.31). It is different from spot height in the sense that it shows the height of the mark but not that of the ground (Singh and Singh 1991). In surveying, *benchmark* acts as a reference point (location), which is used as the base for the measurement of the height of other topographical points. It must be noted that any levelling survey work needs to be started from benchmark.

Based on their nature, methods of determination and utilization, benchmarks are of four main types. These include:

1. GTS benchmark
2. Permanent benchmark

Fig. 4.31 Trigonometrical station, benchmark and spot height

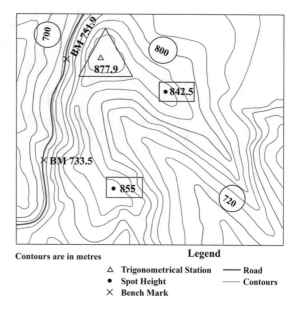

Contours are in metres

Legend

△   Trigonometrical Station   —— Road
●   Spot Height   —— Contours
✕   Bench Mark

3.   Arbitrary benchmark
4.   Temporary benchmark.

(1)   **GTS (Great Trigonometrical Survey) benchmark**: This type of bench-mark is established by the department of Survey of India (SOI) at a long distance all over the country during the time of *Great Triangulation Survey* or *Geodetic surveying*. The values of reduced levels (height or depth from a common datum), their prominent location and the number of the benchmarks are specified in a departmental catalogue of Survey of India.

(2)   **Permanent benchmark**: In this type of benchmark, the reduced level of each point is established by various government departments such as PWD railway, irrigation etc. with reference to the *GTS benchmark*. This benchmark is placed on different permanent objects such as the parapet of a bridge, railway platform, plinth of a building, or culvert and so on. They are also sometimes kept from underground pillar if necessary.

(3)   **Arbitrary benchmark**: Sometimes the levelling work starts choosing any prominent object (like step or plinth of building, permanent rock etc.) as the benchmark whose elevation is assumed arbitrarily. This benchmark is used when we need to carry out small ordinary levelling work or the permanent benchmark is not available at nearby places.

(4)   **Temporary benchmark**: Some levelling survey works require a long period of time to be completed. In such situation, at the end of a day when it is needed to stop the levelling work for that day, any permanent object is selected at which the work is ended and this mark can be used for further survey on the next day.

Fig. 4.32 a Bhola GTS tower near Singur and **b** Semaphore Tower, Parbatichak, Arambagh, Hooghly, West Bengal, India

This type of benchmark is considered as the *temporary benchmark* and is used as reference point to carry out the survey work on the next day and so on.

Trigonometrical Station

The fixed surveying points on the earth's surface that were used as ground stations for the *Great Trigonometrical Survey* or *Triangulation Survey* (an example of *Geodetic Survey*) are known as *trigonometrical station* or triangulation station or triangulation pillar or trig station and sometimes informally as a trig. These stations were typically positioned on hill tops, on temples and in some cases on specially built structures (Fig. 4.32). On maps, these stations are shown by small triangle (Δ) followed by the number, indicating the height of the point above mean sea level (Fig. 4.31). Spot height, Benchmark and Trigonometrical station are collectively called as elevation points.

Form Lines

Forms lines are estimated contours interpolated by eye-sketching between instrumentally fixed contours to represent the approximate configuration of the terrain but not the actual elevations. *Form lines* are commonly shown by broken lines in brown colour (Fig. 4.33) but in some foreign maps, they are shown in the same style as the contours (continuous lines) and thus create confusion. When form lines are shown by broken lines, they are easily distinguished from contours. These lines are drawn without any accurate measure units and are not numbered. Form lines in association with contours are used to indicate the minor details of the topography (like hilly or mountainous country etc.), which are not easily shown by contours in an area of low relief and with large contour interval (Fig. 4.33). In other words, form lines are used where the drawing of contours is not possible but prominent relief feature is there which is lower in height than the contour interval.

Fig. 4.33 Form lines
between contours to show
minor topographic details

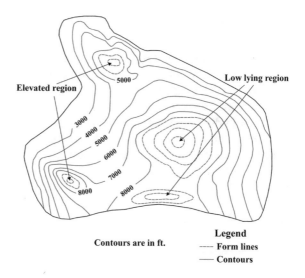

4.11.1.4 Geological Map

Geological maps show the rocks that form the earth's crust, their mode of occurrence and deposition; different geological features like faults, folds etc. of a terrain.

4.11.1.5 Weather and Climatic Map

Weather and climatic maps symbolize various aspects of weather and climate. These maps depict the geographic distribution of the mean monthly or annual values of weather and climatic variables including the temperature, relative humidity, precipitation (rainfall), amount of insolation, cloudiness, atmospheric pressure, direction and velocity of wind etc. Maps showing daily weather conditions are called as daily weather maps and the maps showing the average weather conditions for a period of 10 or more years are known as climatic maps.

4.11.1.6 Soil Map

Maps showing the spatial distribution of different types of soils and their properties are known as *soil maps*.

4.11.1.7 Vegetation Map

Vegetation maps show the spatial distribution of types of vegetation and their characteristics.

4.11.2   Cultural Maps

Maps showing the cultural patterns (i.e. transport network, settlements, monuments etc.) designed by the man over the surface of the earth are called as *cultural maps*. These maps are also used to represent certain details like population characteristics, economic conditions, occupational structure, literacy, social and religious composition etc. in different regions. On small-scale maps, in particular, it is difficult to depict all manmade features without losing legibility. Thus, there is the need to use different types of maps to represent various types of manmade cultural features. Major cultural maps include:

4.11.2.1   Population Map

The *population maps* represent the characteristics of population such as the distribution, density and growth of population, age and sex composition, linguistic and religious composition, occupational structure of the population etc. over an area. Population maps play a very significant role in the proper planning for the solution of different problems and development of an area.

4.11.2.2   Economic Map

An *economic map* illustrates the type and conditions of economic activities present in an area. These maps show the distribution and production of important agricultural crops, minerals and industrial goods, location of industries and markets and their connection with different modes of transportation and communication. They are also concerned with unemployment, energy resource consumption, transport of goods and materials, trade, banking and commerce, economic status etc. of a country, a state or any other region.

4.11.2.3   Political Map

These maps show the state and national boundaries of administrative divisions, locations of different cities, headquarters and capitals of an area like country, state, district etc. They become helpful to the administrative machinery for undertaking different planning and proper management of the concerned administrative area. World Atlas map is an important example of *political map*.

4.11.2.4   Military Map

Military maps record strategic points, routes and battle plans etc.

4.11.2.5   Historical Map

Maps showing the past events are known as *historical maps.*

4.11.2.6   Social Map

Maps that depict the social organisms and aspects such as tribes and races, their languages, religions, castes etc. are called *social maps.*

4.11.2.7   Transportation Map

These maps show the details of different types of roads including major and minor highways, rail lines, the location of railway stations and airports etc. in an area considering the political boundaries and labels. They also show the location of towns and cities, other points of interest like parks, temples, churches and monuments etc. On map, major highways are generally red in colour and wider, whereas minor roads are comparatively lighter and narrower. Ex: National highways, district roads, railway lines etc. are mapped together in *transportation map.*

4.12   Techniques for the Study of Spatial Patterns of Distribution of Elements (Distribution Map)

The *distributional maps* represent the pattern of spatial distribution of any one element based on some specific statistical (or geographical) data. All the maps representing the spatial distribution of various objects or items may be grouped under the head '*the distribution maps*'. These maps display one characteristic feature of a certain area, even ignoring the exact location of the object if necessary. The element or item may be natural like, rainfall, temperature, flora and fauna etc. or it may be cultural like number and density of population, agricultural or industrial products etc. Based on the items, the distribution maps are of different types like—(a) Climatic maps (Isotherm, Isobar, Isohyet maps etc.), (b) Population maps, (c) Crop maps, (d) Mineral maps and (e) Industrial maps etc. In distribution maps, the data may be represented using different colours, symbols, dots, regular lines, shading, bars, circles etc. Distribution maps may be of qualitative or quantitative in nature. When a distribution map becomes diagrammatic, it may be called as *Cartogram.*

Based on the methods of their construction, distribution maps may be classified into different types. It is not possible to employ anyone of the methods for all types of distribution maps. Only one method may be applied to one or two types of the maps. Major types of distribution maps are as follows-

(1)   Chorochromatic map
(2)   Choroschematic map
(3)   Choropleth map
(4)   Dasymetric map
(5)   Isarithmic map (Isometric map and Isopleth map)
(6)   Dot map
(7)   Flow map
(8)   Diagramatic map.

4.12.1 Chorochromatic Map (Colour or Tint Method)

Chorochromatic maps are important qualitative areal distributional (thematic) maps portraying categorical or nominal data using different colours (choro means area and chromos means colour) (Figs. 4.34 and 4.35). These are also known as colour-patch maps. These maps are basically qualitative because the mapping technique does not consider any numerical data.

4.12.1.1   Methods and Principles of Construction

1.   In this technique, the spatial patterns of various features are first outlined and then different colours (or tints) are used to represent the spatial distribution of these features on the map. For example, on a relief map, brown, yellow, green and blue colours are used to depict the mountains, plateaux, plains and seas respectively. Similarly, in a landuse map, forests, agricultural lands and settlement areas are shown by green, yellow and red colours, respectively (Figs. 4.34 and 4.35).
2.   The tint variation of the same colour may effectively be used to portray the spatial variation of the characteristics of a single feature or element. For example, in a vegetation map, tint variation of the green colour (dark shade to light shade) can be used to differentiate evergreen, deciduous, coniferous, desert and other types of vegetations. Similarly, dark shade of blue colour is used to represent the deep sea whereas light shade of blue colour is used to represent the shallow sea. This technique may also be used in relief maps to show different elevations. Tinting method may also be called layering method.
3.   The selection of colours used to show the spatial distribution of various features should be as far as practicable, conventional, realistic and iconic.
4.   A chorochromatic map must include a precisely defined legend of colours indicating the type of features showing on the map (Figs. 4.34 and 4.35).

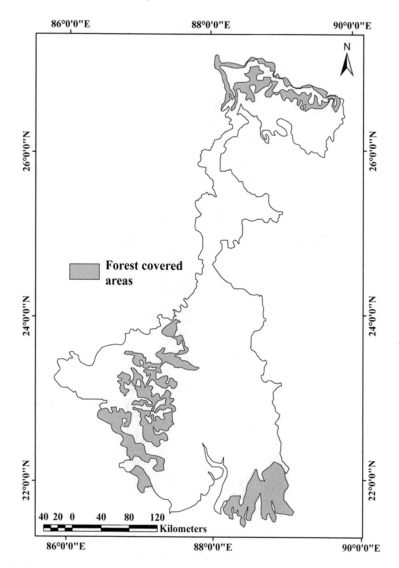

Fig. 4.34 Simple chorochromatic map showing the spatial distribution of forest-covered areas in West Bengal (*Source* NATMO MAPS, DST)

4.12.1.2 Types of Chorochromatic Maps

Chorochromatic maps may be simple or compound in nature:

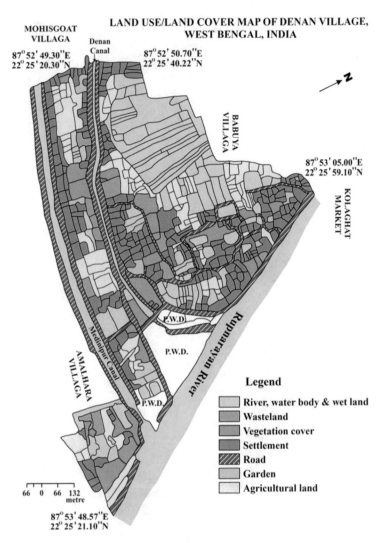

Fig. 4.35 Compound chorochromatic map showing the general land use pattern of Denan village, Purba Medinipur, West Bengal (*Source* Field survey)

Simple Chorochromatic Maps

Simple chorochromatic maps depict a single feature with spatially sporadic or patchy occurrence. The selected feature is represented by one colour pattern without dividing it into different sub-categories.

Example: Maps showing the spatial distribution of coal fields or forest areas in a country or state (Fig. 4.34).

Compound Chorochromatic Maps

Compound chorochromatic maps represent the phenomena in which types and different sub-categories can be identified.

Example: Maps showing the spatial distribution of various types of rocks, soil, vegetation etc. In a soil map, all different kinds of soil are portrayed on the same map with the help of different colours. In a land use/land cover map, various types of landuse features are depicted using different colours (Fig. 4.35).

4.12.1.3   Uses of Chorochromatic Maps

1. Chorochromatic maps are very helpful to represent a wide variety of nominal or discrete data. The combination of various colours facilitates to easily recognize the pattern of spatial distribution of different features.
2. Rock types, relief forms, soil groups, vegetation types, landuse types etc. can effectively be depicted in chorochromatic map.
3. It may advantageously be utilized to show the population density of a region or country. This method may also be used to prepare the isopleth maps by assigning colours in the interval between two isolines.

4.12.1.4   Disadvantages of the Use of Chorochromatic Maps

1. Only one element with or without having different sub-categories can be represented in this map.
2. All other features are completely obliterated in this mapping technique.

4.12.2   Choroschematic or Symbol Map

Choroschematic maps are technically semi-quantitative distributional (thematic) maps showing the spatial distribution pattern of one or more elements (ordinal data) using conventionally selected symbols (letter, pictorial, point, area symbols etc.) (Fig. 4.36). The symbols may be of equal size or varying sizes as the case may be.

4.12.2.1   Methods and Principles of Construction

1. At first, a symbol of uniform size and boldness may be considered as a unit (i.e. a symbol represents a specific quantity) to make the map more useful and easily understandable.
2. When letter symbols are used, the initial letter of an element may be selected as a symbol to represent the same; for example, R for rice, W for wheat, C for

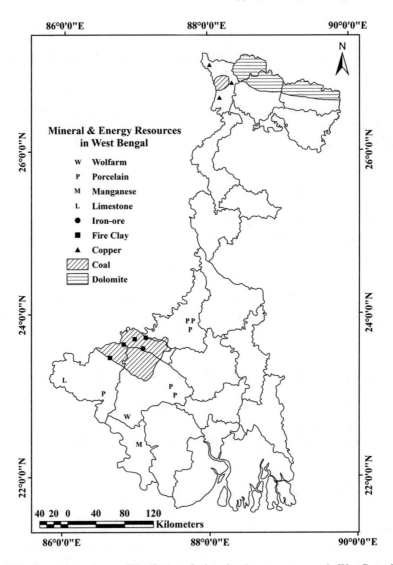

Fig. 4.36 Choroschematic map (Distribution of mineral and energy resources in West Bengal)

cotton etc., in case of crop distribution maps; C for coal, I for Iron ore, M for manganese, G for gold etc., in mineral maps. In case of pictorial forms, the symbols of railway engines and aeroplanes can be used for their representation (Fig. 4.36).

3. When only one element is shown on the map, the required number of symbols may be calculated with reference to the quantity to be represented. Then the symbols are drawn carefully so that they become uniformly distributed over the respective areas.

4. Sometimes, each element may be represented by a symbol of varying size to emphasize the relative importance of different areas with respective to different elements shown on the map. The size of the symbols varies in proportion to the quantity represented, i.e. the height of the symbols especially of letter symbols roughly corresponds to the square root of the quantity. Cressey has used this method in his China's Geographic Foundations to show the distribution of minerals (Singh and Singh 1991).

4.12.2.2 Advantages and Disadvantages of Choroschematic Map

Advantages and disadvantages of the use of *Choroschematic map* are:

Advantages:

1. Spatial distribution of many elements can be shown together on a single map.
2. Such a map may be easily commensurable or proportionate, i.e. measurable by the same standard.

Disadvantages:

1. Size and boldness of the symbols require to be equal and should be uniformly distributed over the areas.
2. Sometimes the actual quantity of an element can't be represented by exact (equivalent) number of symbols. For example, if each symbol represents a quantity of 5,000 units and the actual quantity of an element is 23,500 units. Hence, the number of symbols will be $\frac{23,500}{5,000} = 4.7$. But the drawing of such number of symbols is not possible. We have to consider the number of symbols as 5 (round figure).
3. In case of letter symbols, though represented with the same height, some of the letters may appear larger than others. For instance, M showing manganese may appear large than I showing iron ore in mineral distribution map.

4.12.2.3 Difference Between Chorochromatic Map and Choroschematic Map

Major differences between *chorochromatic* and *choroschematic map* are:

Chorochromatic map	Choroschematic map
1. Qualitative areal distributional map	1. Semi-quantitative distributional map
2. Spatial patterns of various features (nominal data) are shown using different colours or the tint variation of the same colour	2. Spatial distribution pattern of one or more elements (ordinal data) are shown using different symbols (letter, pictorial symbols etc.)
3. Only one element with or without having different sub-categories can be displayed, other features are completely obliterated	3. Spatial distribution of many elements can be shown together

(continued)

(continued)

Chorochromatic map	Choroschematic map
4. **Example**: Soil map, vegetation map, land use/land cover map etc.	4. **Example**: Crop distribution map, mineral distribution map etc.

4.12.3   Choropleth Map

The *choropleth map* may be considered as the chief tool of the human geographers for quantitative treatment of the distributional aspects of population. The literal meaning of choropleth is 'Quantity in Area' (Greek words 'choros' meaning thereby place and 'plethron' meaning thereby measure). Choropleth maps are important quantitative areal maps showing the pattern of spatial distribution of density or intensity of an element with the help of a system of colour or graded shading (Fig. 4.37a, b). This technique is commonly used to show the density or average measures or percentages of different elements, i.e. population density, cropping intensity, sex ratio (male–female ratio), percentage of agricultural land to total land, percentage of irrigated land to total agricultural land etc. within a geographic area.

4.12.3.1   Principles of Construction

1. The basic principle of choroplething is that the intensity of shading (of either line or colour) is directly proportional to the intensity or density of the elements. The lighter shades show lower intensities or densities whereas the deeper or darker shades show higher intensities or densities (Fig. 4.37a, b).
2. The shading generally follows the administrative boundaries (Fig. 4.37a, b), because the very data are available in reference to them.
3. The mapping technique is based on the assumption that population density is homogeneous within the boundary of the areal unit.

4.12.3.2   Methods of Construction

1. For the construction of choropleth map, at first, the intensity, density or percentage values of the given element are arranged in ascending or descending order. If data are not given in these formats then, they are calculated from the given absolute values against each administrative unit (Tables 4.8 and 4.10).
2. The range (maximum value − minimum value) within the values (either given or calculated) is identified, and the number of classes or categories is also selected to represent very high, high, medium, low and very low concentrating nature of the element.

3. The interval between successive categories may be chosen theoretically following the formula:

$$\text{Interval} = \frac{\text{Range of values}}{\text{No. of categories}} \qquad (4.18)$$

Sometimes, the interval is chosen based on experience rather than following the theoretical formula to avoid the empty category or concentration of large number of values in one or two category/ies.

4. The intensity, density or percentage values of all the administrative units are then grouped against the selected categories (Tables 4.9 and 4.11).

5. Suitable shades (of either line or colour) are used against each administrative unit to depict the chosen categories. A Choropleth map must include a precisely defined index or legend of shades indicating the nature of distribution of the element showing on the map (Fig. 4.37a, b).

4.12.3.3   Advantages and Disadvantages of Choropleth Map

Major advantages and disadvantages of use of choropleth map are discussed below:

Advantages:

1. Simple visual representation and impression of spatial variation of distribution of elements using shades of line or colour.

2. Easy to understand and interpret not only to the geographers or literate people but also to the common illiterate people using the index or legend key.

Disadvantages:

Although choropleth maps provide an excellent visual impression of spatial variation of distribution of elements but there are certain disadvantages of using this method:

1. Reading of actual values or figures is not possible from the selected categories.

2. It provides wrong impression of sudden change of values at the boundaries of the administrative units. As the shading follows the administrative boundary, so the boundary of different shading may not follow the exact boundary of variation of the element on the real ground. Because, on surface, the spatial distribution of an element (say population density or cropping intensity) never changes abruptly rather it changes gradually with distance.

3. Spatial variations within each administrative area are hidden. The element may not be uniformly distributed over the entire administrative area; so some very small areas showing a higher density will be obliterated by areas of moderate density.

4. Moreover, little consideration is possible regarding unused or waste lands like, deserts, rugged and rocky areas, mountains, hills etc. All the waste lands or unused lands on the surface disappear beneath the shading in map.

4.12.3.4   Representation of Population Density in Choropleth Map

Choropleth maps are widely and popularly used by the human geographers to represent the spatial variation of distribution of *population density* over geographic areas.

Table 4.8 shows the *population density* of different districts of West Bengal, India for the census year 2011. Here, the highest and lowest population densities are found in Kolkata (24,306/km^2) and Kalimpong (239/km^2) districts, respectively. As the highest population density is exceptionally high compared with other values, the said formula (Eq. 4.18) has not been followed to calculate the value of range of population density, interval and the number of density categories because it will create some categories having no population density within this range (i.e. empty categories). Thus, the number of categories and their interval has been determined manually based on experience to avoid the empty category/ies (Table 4.9).

4.12.3.5   Representation of Cropping Intensity in Choropleth Map

Cropping intensity can be defined as the frequency of occurrence of crops in agricultural lands in a given region over a given time period. Increasing the cropping intensity is one of the important methods to increase the agricultural production in a region.

Cropping intensity is the ratio of *Gross Cropped Area (GCA)* to *Net Cropped Area (NCA)*, which is multiplied by 100 and represented in percentage (%).

$$\textbf{Cropping Intensity (CI)} = \frac{\text{Gross Cropped Area (GCA)}}{\text{Net Cropped Area (NCA)}} \times 100 \qquad (4.19)$$

Single cropping intensity means the agricultural lands are cultivated once in a year, double cropping intensity means the agricultural lands are cultivated twice in a year, multiple cropping intensity means the agricultural lands are cultivated three times in a year.

For example, if a region has 100 hectares of agricultural lands, and the lands are cultivated twice in a year, then Net Cropped Area is 100 hectares and Gross Cropped Area is (100 hectares $\times$ 2) = 200 hectares and the

$$\text{Cropping Intensity} = \frac{(100\,\text{hectares} \times 2) = 200\,\text{hectares}}{100\,\text{hectares}} \times 100 = 200\%.$$

Again, if 100 hectares of lands are cultivated in one season and only 50 hectares of lands are cultivated in another season then:

$$\text{Cropping Intensity} = \frac{(100\,\text{hectares} + 50\,\text{hectares}) = 150\,\text{hectares}}{100\,\text{hectares}} \times 100 = 150\%.$$

Table 4.8 Worksheet for choropleth map (Population density of different districts of West Bengal, 2011 census)

Sl. Number	Name of the districts	Area (km^2)	Population (census 2011)	Population density (number of persons/km^2) $\left[\dfrac{\text{Total population}}{\text{Total area}}\right]$
1	Alipurduar[*]	3,136	1,501,983	479
2	Bankura	6,882	3,596,292	523
3	Paschim Bardhaman	1,603.17	2,882,031	1,798
4	Purba Bardhaman	5,432.69	4,835,432	890
5	Birbhum	4,545	3,502,404	771
6	Cooch Behar	3,387	2,819,086	832
7	Darjeeling	2,092.5	1,846,823	883
8	Uttar Dinajpur	3,140	3,007,134	958
9	Dakshin Dinajpur	2,219	1,676,276	755
10	Hooghly	3,149	5,519,145	1,753
11	Howrah	1,467	4,850,029	3,306
12	Jalpaiguri	3,044	3,872,846	1,272
13	Jhargram[*]	3,037.64	1,136,548	374
14	Kolkata	185	4,496,690	24,306
15	Kalimpong[*]	1,054	252,239	239
16	Malda	3,733	3,988,845	1,069
17	Paschim Medinipur	6,308	5,913,457	937
18	Purba Medinipur	4,736	5,095,875	1,076
19	Murshidabad	5,324	7,103,807	1,334
20	Nadia	3,927	5,167,600	1,316
21	North 24 Parganas	4,094	10,009,781	2,445
22	South 24 Parganas	9,960	8,161,961	819
23	Purulia	6,259	2,930,115	468

[*] Was created after the 2011 census (*Source* Census of India)

In Table 4.10, the highest and lowest cropping intensities are found in Hooghly (250%) and Paschim Bardhaman (115%) districts, respectively.

The range of cropping intensity = 250% − 115% = 135%.

If five categories are considered, then interval of each category will be $\frac{135}{5} = 27$.

Table 4.9 Category-wise population density in different districts in West Bengal (2011 census)

Sl. Number	Categories of population density (Number of persons/km^2)	Name of the districts
1	<500	Alipurduar, Jhargram, Kalimpong, Purulia
2	500–1,000	Bankura, Purba Bardhaman, Birbhum, Cooch Behar, Darjeeling, Uttar Dinajpur, Dakshin Dinajpur, Paschim Medinipur, South 24 Parganas
3	1,000–1,500	Jalpaiguri, Malda, Purba Medinipur, Murshidabad, Nadia
4	1,500–2,000	Paschim Bardhaman, Hooghly
5	>2,000	Howrah, Kolkata, North 24 Parganas

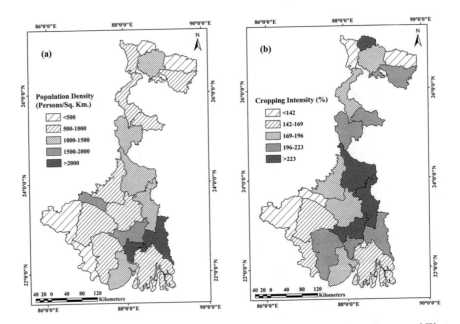

Fig. 4.37 a Population density map of West Bengal (2011), **b** Cropping intensity map of West Bengal (2018–2019)

Therefore, the categories will be like (Table 4.11):

1. <142 (115 + 27)
2. 142–169 (142 + 27)
3. 169–196 (169 + 27)
4. 196–223 (196 + 27)
5. >223.

Table 4.10   Worksheet for cropping intensity map of West Bengal (2018–2019)

Sl. Number	Name of the districts	Gross cropped area (GCA) in hectares	Net cropped area (NCA) in hectares	Cropping intensity (CI) in % = ($\frac{GCA}{NCA} \times 100$)
1	Alipurduar[*]	187,719	135,300	139
2	Bankura	541,194	340,380	159
3	Paschim Bardhaman	63,261	55,010	115
4	Purba Bardhaman	740,072	400,050	185
5	Birbhum	581,559	327,400	178
6	Cooch Behar	547,042	257,000	213
7	Darjeeling	153,106	115,805	132
8	Uttar Dinajpur	512,881	273,200	188
9	Dakshin Dinajpur	388,682	186,900	208
10	Hooghly	527,233	210,895	250
11	Howrah	165,003	80,400	205
12	Jalpaiguri	381,280	203,800	187
13	Jhargram[*]	221,060	139,900	158
14	Kolkata	0	0	0
15	Kalimpong[*]	56,696	23,400	242
16	Malda	467,038	231,800	201
17	Paschim Medinipur	832,316	385,800	216
18	Purba Medinipur	532,773	284,810	187
19	Murshidabad	937,356	399,000	235
20	Nadia	662,573	293,700	226
21	North 24 Parganas	500,385	230,900	217
22	South 24 Parganas	603,658	358,300	168
23	Purulia	369,801	314,507	118
Total		9,972,688	5,248,257	190

[*] Was created after the 2011 census

Source   http://matirkatha.net/wp-content/uploads/2016/07/West-Bengal-District-wise-Cultivated-Area-Gross-cropped-are-cropping-intensity-2018-19.pdf

Table 4.11   Category-wise cropping intensity in different districts in West Bengal (2018–2019)

Sl. Number	Categories of cropping intensity (%)	Name of the districts
1	<142	Alipurduar, Paschim Bardhaman, Darjeeling, Purulia
2	142–169	Bankura, Jhargram, South 24 Parganas
3	169–196	Purba Bardhaman, Birbhum, Uttar Dinajpur, Jalpaiguri, Purba Medinipur
4	196–223	Cooch Behar, Dakshin Dinajpur, Howrah, Malda, Paschim Medinipur, North 24 Parganas
5	>223	Hooghly, Kalimpong, Murshidabad, Nadia

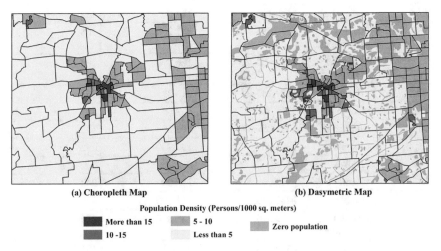

(a) Choropleth Map   (b) Dasymetric Map

Population Density (Persons/1000 sq. meters)

■ More than 15	■ 5 - 10	■ Zero population
■ 10 -15	■ Less than 5	

Fig. 4.38   Visual difference between the Choropleth map (**a**) and the Dasymetric map (**b**) (Dasymetric map shows the exclusion areas of zero population)

4.12.4   Dasymetric Map

The *dasymetric map* is a technique of thematic mapping in which areal symbols are used for spatial classification of volumetric data. The concept of dasymetric map was first described and developed by Semenov-Tyan-Shansky in 1911, and it was popularized by J.K. Wright in 1936. The term '*dasymetric*' was derived from the Greek translation for 'measuring density' ('dasys' means dense and 'metreo' means measure). Semenov-Tyan-Shansky defined dasymetric maps as maps 'on which population density, irrespective of any administrative boundaries, is shown as it is distributed in reality, i.e. by natural spots of concentration and rarefaction' (Fig. 4.38b).

Dasymetric mapping is an area-based cartographic technique in which cartographers have attempted to minimize the defects and improve upon choropleth mapping by incorporating statistics and ancillary geographical information such as remotely

sensed data, transport network, topographic maps and land cover datasets to esti-
mate the true population distribution (Fig. 4.38a, b). This method can depict the
spatial variations of population density more realistically on map as boundaries are
adjusted to conform to known homogeneous regions and are not confined to administrative boundaries, i.e. administrative boundaries are not taken into consideration
in dasymetric map (Fig. 4.38a, b).

For instance, superimposition of aerial photos with the studied areas will partially
be helpful to identify the unpopulated areas which should, then, be eliminated from
the total land area during the time of calculation of population density.

4.12.4.1 Methods and Principles of Construction

1. *Dasymetric mapping* is a cartographic technique, in which the actual administrative areas are necessarily divided into smaller areas, onto which the socio-
   demographic attribute (say population) is averaged for the calculation of the rate,
   like population density. It uses the boundaries that divide the area into different
   homogeneous zones in purpose of better representation of the distribution of
   population.
2. This technique is explicitly based on the fact that population may not be
   uniformly distributed over the entire administrative area rather certain areas
   are populated, while others are not (Wright 1936). For example, in dense urban
   areas, composed of smaller areal units, the population density may be homogeneous, but population densities differ significantly outside of the urban centre
   within the block group boundary.

3. Considering a compromise between choropleth and isopleth maps, dasymetric
   mapping uses standardized data and takes into account the exact changing densities of population within the boundaries of the map. To serve this purpose, additional (ancillary) information is collected within the area of study and the cartographer steps statistical data by incorporating the additional information. Proper
   execution of this technique makes the dasymetric map superior to choropleth
   map in conveying statistical data within an area.

A simple and common example is that different types of uninhabited regions like
water bodies, ice-covered areas, government-owned land etc. are excluded from the
choropleth districts and thus they appear vacant in the dasymetric map (Fig. 4.38b).

4.12.4.2 Uses of Dasymetric Map

Cartographers use *dasymetric map* to represent the density of population over other
techniques because it displays the data more realistically on map. Dasymetric maps
are not so popular because of the limited alternatives for making them with automated
computerized tools like geographic information system. Although choropleth maps
are popularly used in various fields but the uses of dasymetric maps are increasing
in developing sectors, like conservation and sustainable development.

4.12.4.3   Difference Between Choropleth Map and Dasymetric Map

Major differences between *choropleth map* and *dasymetric map* are:

Choropleth map	Dasymetric map
1. This technique is generally used to show the density or intensity of different elements (say population density) using a system of colour or graded shading	1. The defects of choropleth map are minimized and quality is improved by incorporating ancillary geographical information like remotely sensed data, topographical maps etc.
2. System of colour or graded shading generally follows the administrative boundaries (district boundaries, state boundaries etc.)	2. Population density, irrespective of any administrative boundaries, is shown as it is distributed in reality
3. It is based on the assumption that population density is homogeneous within the boundary of the areal unit	3. It believes that certain areas are populated, while others are not, i.e. it considers the natural spots of concentration and rarefaction
4. Drastic change of population densities occurs along the boundaries of the administrative units	4. It takes into account the exact changing densities of population within the boundaries of the map
5. Different uninhabited regions like water bodies, government-owned lands etc. are also shaded within administrative units	5. Uninhabited regions like water bodies, government-owned land etc. remain vacant in the dasymetric map
6. Choropleth maps are widely used by the cartographer in different fields because of their simplicity and uncomplicated construction methods	6. Though dasymetric maps are superior to choropleth maps but are not so popular because of the limited alternatives for making them with automated computerized tools like geographic information system

4.12.5 Isarithmic Map (Isometric Map and Isopleth Map)

Isarithmic map is an important type of thematic (distributional) map that shows various continuous geographical variables using line and/or region symbols to join the places having equal value. Generally, these maps are used for the visualization of continuous data sets with the help of graded shading, colour, particularly tint of colour and value (Figs. 4.39b and 4.40).

Isarithmic mapping normally uses control points having two kinds of data: (a) the true point data and (2) the conceptual point data to draw the isolines (*Isolines* are lines connecting places of equal value). True point data are the actual values collected or recorded at certain locations whereas the data recorded over an area are considered as conceptual point data.

4.12.5.1   Types of Isarithmic Map

Based on the kinds of data used, Isarithmic maps can be divided into two main types:

1. **Isometric Map**: If the points indicate actual true data, the values of the isolines can be explained as exact and the map is called an *isometric map*. This map is constructed from true point data recorded at sampled locations and shows the distribution of actual or derived quantities. For example, Temperature map is constructed using the temperature values recorded at individual weather stations across a country or region. Rainfall map, height map, toxic level map etc. are also the examples of Isometric map.
2. **Isopleth Map**: When the data are areal, the isolines cannot be interpreted as exact and the map is called an *isopleth map*. This map is constructed from conceptual point data based on quantities that cannot exist at point locations and are expected to have a larger intrinsic positional error. For example, demographic trend map is generated from the conceptual point data at the centroid of enumeration unit.

Therefore, Isopleth maps and Isometric maps are the examples of Isarithmic maps and are believed to be very similar. The only difference between these two maps is the type of data used and the precision of the data representation; no distinction exists on account of the kinds of cartographic symbols. Thus, isopleth maps are widely used for the representation of the continuous geographical variables with clearly observable numerical values, i.e. Isarithmic map and Isopleth map are used synonymously in geographical perspectives.

The *isopleth map* is considered as the basic tool of the climatologists for quantitative analysis and representation of the spatial distribution of climatic parameters. The term Isopleth is a combination of two Greek words 'Isos' meaning thereby same or equal and 'Plethron' meaning thereby measure. Thus isopleths are the lines joining places of equal values in the form of quantity, intensity, density etc.

4.12.5.2   Commonly Used Isolines or Isopleths

Some commonly used *isolines* or *isopleths* are:

- **Contours**: lines of equal elevation
- **Isogons**: lines of equal angle or direction (such as magnetic direction)
- **Isogonals**: lines of equal magnetic declination
- **Isopors**: lines of equal annual change in magnetic declination
- **Isotherms**: lines of equal temperature
- **Isobars**: lines of equal atmospheric pressure
- **Isohyets**: lines of equal rainfall
- **Isohels**: lines of equal amount of sunshine or solar radiation
- **Isosteres**: lines of equal atmospheric density
- **Isohumes**: lines of equal relative humidity

- **Isonephs**: lines of equal cloud cover or cloudiness
- **Isotachs**: lines of equal wind speed
- **Isopachs**: lines of equal atmospheric thickness
- **Isogeotherms**: lines of equal annual mean temperature
- **Isocheims**: lines of equal mean winter temperature
- **Isotheres**: lines of equal mean summer temperature
- **Isodrosotherms**: is a line of equal dew point
- **Isochalazs**: lines equal frequency of hail storms
- **Isobronts**: lines of simultaneous occurrence of given phase of thunderstorm activity
- **Isochasms**: lines of equal recurrence of auroras
- **Isoplats**: lines of equal acidity as in acid rain (precipitation)
- **Isochrones**: lines of equal travel time
- **Isodopes**: lines of equal Doppler Radar velocity
- **Isodoses**: lines of equal intensity of radiation
- **Isodapanes**: lines of equal total transport costs
- **Isotims**: lines of equal transport costs separately for raw materials or finished goods
- **Isophenes**: lines of equal time of occurrence of biological events like crops flowering
- **Isobaths**: lines of equal depth under water
- **Isobathytherms**: lines showing depths of water with equal temperature
- **Isohalines**: lines of equal ocean salinity
- **Isopycnals**: surfaces of equal water density.

4.12.5.3 Methods and Principles of Construction

1. Generally, Isopleth maps are generated either by interpolating from raster data or from the values recorded at different sample locations. For example, in case of isopleth map showing the spatial distribution of temperature, the recorded values of temperature at individual weather stations across a region or country are at first classified into different groups like, <40, 40–50, 50–60, 60–70, 70–80, 80–90 °F etc. (Fig. 4.39) based on the minimum and maximum values of temperature and the selected interval.

2. Then, required numbers of isotherms (lines of equal temperature) are drawn upon the map to show various temperature zones and value, colour or saturation is added to enhance the quality and interpretability of the map. Generally, interval of 5, 10 or 20 is expected to be ideal. The value of the Isolines should be mentioned on either side of the lines or in the middle by breaking the lines (Fig. 4.39). The isolines should not be too dense and the interval of the isolines depends on: 1. Purpose of the map-making, 2. Scale of the map, 3. Reliability of data, 4. Data density, 5. Nature of the data distribution etc.

3. Generally, the data for the Isopleth map are collected from scattered sample locations, having no data between these sample locations. This is the reason

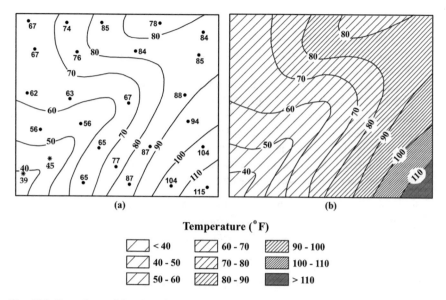

Fig. 4.39   Procedures of drawing of Isopleth map (Isotherms in this sample area)

why interpolation is useful in Isoplething. Interpolation is a technique that allows obtaining the values of intermediate points derived from the existing observed values (control points), i.e. unknown values are estimated from the known values. Method of interpolation assumes that all the intermediate values are evenly distributed between these control points. A few distinguished interpolation methods are:

1. Inverse Distance Weighting
2. Kriging
3. Triangulation.

The correct point of drawing an Isoline can be determined using the formula given below:

$$\text{Exact point of isoline} = \frac{\text{Distance between two control points}}{\text{Difference between two values of the corresponding points}} \times \text{Interval}$$

$$(4.20)$$

Here, interval is the difference between the actual value of a point on the map and the interpolated value. For example, in the Isotherm map (Fig. 4.39a), two places show 39 and 45 °F temperature and we want to draw 40 °F Isotherm. Assume that the distance between two observed points is 1.2 cm or 12 mm, and the difference between 39 and 45 °F is 6 °F. Now, 40 °F is 1 point away from 39 °F and 5 points behind 45 °F, therefore, the exact point of 40 °F temperature will be plotted 2 mm away from 39 °F temperature or 10 mm ahead of 45 °F temperature.

4. Isopleth maps can be of two forms:

    First, lines of equal value are drawn, which indicate that all the values lying on one side of the *'isoline'* are lower than it while the values on the other side are larger than it (Fig. 4.39a).

    Second, sometimes the zones of similar value are represented by similar patterns or colours (Figs. 4.39b and 4.40).

5. In case of interpreting an isopleth map, attention should be paid on the zones lying between the lines. The gap between the isolines of specific value helps the reader to understand the rate of change of the value from one line to the next and can easily be measured. Lines that are close to each other indicate a steep gradient while the lines that are widely spaced are characterized by gentle gradient.

In Fig. 4.40, the distribution of annual rainfall (in cm) of West Bengal is shown using isopleth technique irrespective of the district boundaries. Different shades have been used to represent different rainfall zones.

4.12.5.4 Advantages and Disadvantages of Use of Isarithmic Map

Major advantages and disadvantages of using an isarithmic (isopleth) map are:

Advantages:

- The spatial distribution of a smooth continuous data can effectively be represented in isopleth map. For example, temperature exists at every point on the earth's surface (is continuous) and does not change suddenly and thus should be mapped using isoplething technique. Rainfall, relief etc. should also be represented in isopleth map for this reason.
- The rate of change (gradient) of the variables can easily be understood in this technique. Wide spacing of isolines indicates the lower rate of change of variables and vice versa.
- The map is independent of administrative or political boundaries and hence most useful to show the actual pattern of distribution of geographical variables (Fig. 4.40).
- Map can easily be adjusted to various levels of scale because of its flexibility.
- It is the best-suited method to show the variables having transitional belt. Because of this, isopleth map is popularly used to show the distribution of rainfall, temperature, air pressure etc.
- Maps can easily be created using computer cartographic tools and techniques.

Disadvantages:

Despite of the above mentioned advantages, this method suffers from several disadvantages:

- This technique requires a large amount of data spreading all over the area under study and the changes are gradual.

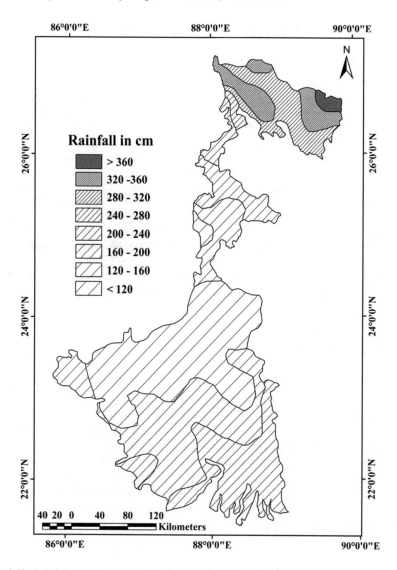

Fig. 4.40 Rainfall zones of West Bengal in Isopleth map (*Source* NATMO MAPS, DST)

- These maps are not suitable for showing discontinuous or 'patchy' distributions of variables.
- The method of isoplething often needs the interpolation technique, which is difficult to use and may not represent the true nature of distribution of the variables.
- In case of narrow transitional belt and abrupt change in data distribution, the isolines lose their importance.

4.12.5.5   Difference Between Choropleth Map and Isopleth Map

Major differences between *choropleth* and *isopleth maps* are as follows:

Choropleth map	Isopleth map
1. The spatial distribution of different geographical variables is shown by shading or using tints of the same colour	1. The spatial distribution of geographical variables is shown with the help of isolines
2. Suitable for discontinuous variables. Continuous variables can also be represented	2. Distribution of continuous variables is represented only
3. Administrative boundary demarcates the data boundary and hence representation of continuous variables may be misleading	3. Data boundary is not defined by administrative boundary rather it is demarcated by data patterns with the help of isolines
4. Colour gradient or shading helps the map interpretation. Deeper the colour, higher the value of the variable and vice versa	4. Spacing of the isolines helps to interpret the map. Closer the isoline, steeper is the distribution of variables and vice versa
5. Mainly used to show population density, cropping intensity, sex ratio etc.	5. Mainly used to show the spatial distribution of altitude, temperature, rainfall etc.

4.12.6   Dot Map

A *dot map* is an important type of thematic map, which uses dots to depict the presence of an element and displays its spatial pattern of distribution. It is a common mapping technique that shows the distribution of phenomena having specific quantities (values) and location using dot symbols and hence considered as quantitative distribution map. Dot maps are especially useful when the data are unevenly and sporadically distributed (like the distribution of rural population) and create a visual impression of concentration of the element being mapped by placing dots in the approximate position in the map. Dot maps are basically suitable for the mapping of the absolute data like population, stocks, crops, minerals etc.; relative figures cannot be represented in this technique.

4.12.6.1   Methods and Principles of Construction

Selection of the Suitable Dot Scale

In case of drawing a dot map, at first a suitable dot scale should be selected, i.e. each dot representing a specific number or quantity of the element. For the selection of suitable scale, three things must be taken into consideration—(a) scale of the map, (b) minimum and maximum values and (c) the type of element to be mapped.

In small-scale map, the value represented by each dot should be so selected that the number of dots may not be too many, because too many dots placed on a small-scale map will create a blurred effect, i.e. the area with low concentration will get a dark appearance. In large-scale map, too few dots will create similar type of problem, i.e. the area with high concentration of will get a light appearance.

The unit of each dot should be selected based on the minimum and maximum values in such a way that, as such as possible, no area is left unrepresented on the map.

The values represented by each dot may also be different depending on the type of element to be shown on the map. For example, the number represented by each dot will be less in case of cattle distribution than that of in the distribution of goat or ship. It is because of the fact that, the number of cattle to be reared in a certain region will definitely be less than the number of goat or sheep in the same region. Thus, if one dot represents 200 cattle, then in case of goat or sheep, one dot may represent 300 or 400 numbers.

The number of dots (n) corresponding to an administrative unit is directly proportional to the quantity (q) to be represented and empirically it can be expressed as:

$$n \propto q$$
$$n = k.q \tag{4.21}$$

where, k = constant of proportionality.

After calculating the required number of dots to represent a specific quantity, dots should be placed properly within the respective administrative units.

Selection of Size of Dots

The size of the dots should be chosen judiciously to make the dot map effective and successful. Dot size should neither be so large that the whole map appears congested nor it should be so small that the areas with greater concentration of the given element look vacant. The general principle of selecting dot size is that in areas of highest concentration, the dots should just touch one another. Some scholars opine that in areas of maximum concentration, the coalescence of the dots should just start. Again, few scholars are of the opinion that the size of the dot should be selected such that they can just be counted in areas of highest concentration.

Most important principle of dot method is that all the dots should be uniform in size because each dot represents a specific value or number. Uniform size of the dots cannot possibly be drawn by ordinary pens. For this reason, special types of nibs like dotting nibs, ball point nibs, Le Roy, Payzant etc. may be used for depicting the dots.

Spacing or Placing the Dots

Great care and precaution is required for proper spacing of the dots on the map. The dots should be placed exactly in the same way as the concerned element is actually distributed over the surface, i.e. the areas of high concentration are characterized by large number of dots while lesser number of dots is placed in low concentration areas. No dot should be placed over negative areas like water bodies, rugged infertile topography, marshy lands, dense forest areas etc. having no concentration of the concerned element. Other accessory maps like relief, climate, drainage, soil and vegetation maps should be consulted properly to identify the location of these negative areas. The scale of the map is very important for accurate marking of the positions of negative areas on map. For example, on large-scale map, these areas can be marked easily and accurately compared with the small-scale map. Different negative areas should be lightly marked with pencil and after placing of dots on the map, the pencil marks should be erased properly.

Under even distribution system, dots are placed evenly at equal distances considering three points—(1) dots should not be placed in straight rows, (2) dot lines should not be parallel to the boundaries of the administrative units and (3) continuity should be maintained with the dots placed in adjacent administrative units. Under uneven distribution system, each dot is put at the centre of gravity of distribution of the element. For instance, if one dot is needed to represent the population of a city, the dot will be put at the area of maximum concentration of population in the concerned city. In rural areas, dots are put in areas that are actually inhabited. This method of placing the dots is more accurate than the former but detail geographical knowledge (thorough understanding of topographical, climatic, soil, vegetation cover etc.) of the concerned area is required for using this technique.

Table 4.12 represents the number of rural population in different districts in West Bengal (2011 census) and the spatial distribution of this rural population is depicted in Fig. 4.41 using dots.

Multiple Dot Method

Multiple dot method is very important to make the dot map more clear, accurate and diversified. This method is mainly used in two situations:

First, sometimes, dots of different sizes, each dot representing a certain number or quantity are used to show the distributional pattern of various elements accurately. This method becomes essential and helpful to represent the spatial distribution of elements in some specific circumstances. For example, in case of population distribution, when a very densely populated region is close to a very sparsely populated region, no single dot unit will properly depict the true picture of the distribution in both the regions. If each dot represents a small number, then large number of dots is needed to represent the population in the dense region causing the problem of overcrowding of the dots in the densely populated region. Again, if each dot represents a large number, then least number of dots is required to represent the population in the sparse region which will not depict the true nature of distribution in the sparsely populated region as the population will appear to be concentrated at few points. Under this circumstance, dots of different sizes may be used to show exceptionally dissimilar number of populations. Multiple dot methods can effectively be used to depict the rural and urban population of a region on a single map. Here, small dots can be used to represent the populations of rural areas or small towns, medium dots can be used to represent the populations of medium towns, and large dots can be used to represent the populations of large cities. For example, if we want to show the

Table 4.12 Calculations for dot map (Status of rural population in different districts of West Bengal, census 2011)

Sl. Number	Name of the districts	Rural population	Scale selection	Number of dots
1	Alipurduar*	1,183,704	1 dot to 50,000 rural population	24
2	Bankura	3,296,901		66
3	Paschim Bardhaman	530,077		11
4	Purba Bardhaman	4,109,169		82
5	Birbhum	3,052,956		61
6	Cooch Behar	2,529,652		51
7	Darjeeling	923,410		18
8	Uttar Dinajpur	2,644,906		53
9	Dakshin Dinajpur	1,439,981		29
10	Hooghly	3,390,646		68
11	Howrah	1,775,885		36
12	Jalpaiguri	1,628,791		33
13	Jhargram*	1,096,996		22
14	Kolkata	0		0
15	Kalimpong*	195,450		4
16	Malda	3,447,185		69
17	Paschim Medinipur	4,093,775		82
18	Purba Medinipur	4,503,161		90
19	Murshidabad	5,703,115		114
20	Nadia	3,728,727		75
21	North 24 Parganas	4,277,619		86
22	South 24 Parganas	6,074,188		121
23	Purulia	2,556,801		51

* Was created after the 2011 census

Source Primary census abstract and district census report, 2011

population of Kolkata, Howrah and the surrounding rural areas (villages), then size of dot will be small for the surrounding rural areas (villages), medium for Howrah and large for Kolkata.

Second, multiple coloured dots are used for the representation of two or more elements on a single map. For example, different crops that do not overlap each other like wheat and rice, sugar cane and sugar-beet, cotton and jute etc. can easily be shown on map using multiple coloured dots.

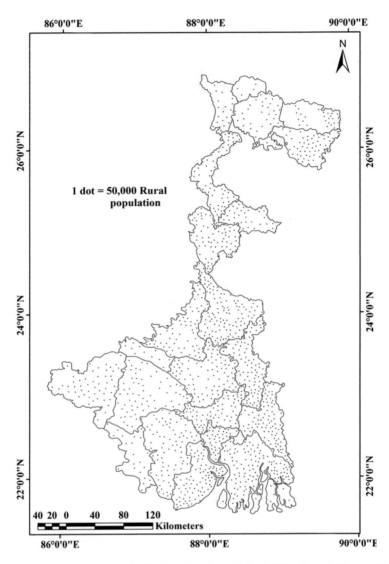

Fig. 4.41 Dot map showing the distribution of rural population in West Bengal (*Source* Primary census abstract and district census report, 2011)

Advantages and Disadvantages of Dot Maps

Advantages:

1.   Spatial pattern and variation of distribution of elements can effectively be shown in this technique.
2.   This is an important method to show the absolute figures and very easy to construct and understand. We can get an idea of the values depicted on the map simply by counting the number of dots.
3.   It provides better visual impression than other methods of distributional maps.
4.   Dot map can easily be converted into choropleth or isopleth map.
5.   More than one component can effectively be represented on a single map by using multiple dot method.

Disadvantages:

1.   This technique is suitable for the representation of absolute figures only. Different relative values like population density, cropping intensity, proportion of rural population to total population etc. cannot be shown by this technique.
2.   It requires a large amount of initial information and detail geographical knowledge (like relief, drainage, soil, vegetation characteristics etc.) of the concerned region to correctly locate the negative areas which are essential for spacing the dots.
3.   It loses much of its importance because of unavailability of data for small divisions of the concerned area.
4.   It is very difficult to count a large number of dots on the map in order to know the actual value represented.
5.   More possibility of making errors by the cartographers in spacing the dots on a dot map.

4.12.7   Flow Map

Flow map is an important type of *dynamic thematic maps*, which are popularly used in *cartography* to show the nature or degree of concentration and movement of people, goods, information etc. between and among places or areas along different routes of transportation and communication. The variations in the quantities of movement of different objects or information are conveyed by the lines of varying widths considering the actual route of movement on a map. Flow maps are commonly used to represent the movement of buses, trains and other modes of transportation; movement of people (migration); movement of goods (retail and household goods); circulation of newspaper and other information; flow of water in rivers and canals etc. Much attention is paid on the following things while drawing a flow map:

(a) Name and type of good or object that is flowing or moving or migrating; like vehicles, commodities, people, water in river etc.
(b) The direction of the flow or movement, i.e. the origin and destination of the same.
(c) The quantity (how much) of flow or movement or transfer from one place to another.
(d) General information about how it is flowing or moving.

4.12.7.1 Methods of Construction

Steps followed sequentially for the drawing of a flow map are:

1. At first, a map of the concerned area depicting the desired transport routes along with the important connecting stations or nodes (the place of convergence of goods or traffic) is prepared. In case of water flow in river or canal, the drainage map is taken into consideration. The data regarding the flow of goods, services; movements of vehicles, people; flow of water etc. from the origin to the destination are compiled.
2. A suitable scale is selected for the representation of the above mentioned data.
3. For easy drawing of the flow map, a line of uniform thickness may be considered as a unit and requisite number of such lines may be placed closely parallel to each other to represent the flow or movement of various numbers or quantities of the given data. More the number or quantity greater the thickness of the line. Actual route and direction of flow or movement are truly maintained in this map.
4. When the quantity of goods or traffic along a particular route of movement is a hundred or thousand times greater than along other routes, then the thickness of the corresponding line may not be proportional to the quantity represented. In such situation, an index of line/s with decreasing thickness need to be made and shown in a corner of the map in which the thickest line will correspond to the maximum quantity of flow or movement.
5. In case, the quantity of goods or traffic is very highly concentrated on a particular station or node, a separate flow map of that particular station may be prepared on such a large scale that the high concentration may be shown clearly, and it may be placed in the main map as an inset map.

Table 4.13 represents the number of local trains connecting between selected stations (daily) in West Bengal, India, and the movement of these trains is graphically depicted in Fig. 4.42 with the help of flow lines with varying thickness. More width of the flow lines towards Howrah (H) station reveals its more importance compared with other stations (nodes) in this rail network system. Santragachi, Uluberia, Mecheda, Panskura, Kharagpur are also important nodes as depicted. Contrary to this, Bargachia, Amta, Shalimar and Haldia are less important stations as the widths of the flow lines are comparatively less.

Table 4.13 Worksheet for flow map (Number of local trains connecting between selected stations in West Bengal, India)

Line	Number of local trains connecting between the stations (daily)	Total number of local trains moving between stations	Scale selected	Width of the flow line (cm)
Howrah (H)–Midnapur (MD)	12	12	(1 cm width to 60 local trains)	0.20
Howrah (H)–Kharagpur (K)	08	12 + 8 = 20		0.33
Howrah (H)–Panskura (P)	22	12 + 8 + 03 + 22 = 45		0.75
Howrah (H)–Haldia (HL)	03	03		0.05
Howrah (H)–Mecheda (M)	04	12 + 8 + 03 + 22 + 4 = 49		0.82
Howrah (H)–Uluberia (U)	03	12 + 8 + 03 + 22 + 4 + 3 = 52		0.87
Howrah (H)–Amta (A)	04	04		0.07
Howrah (H)–Shalimar (S)	05	05		0.08
Howrah (H)–Bargachia (B)	04	04 + 04 = 08		0.13
Howrah (H)–Santragachi (SR)	02	12 + 8 + 03 + 22 + 4 + 3 + 2 = 54		0.90

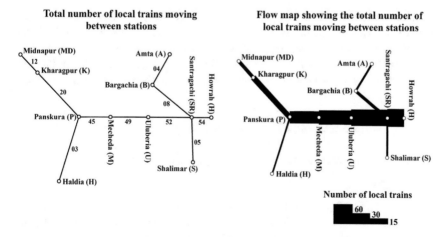

Fig. 4.42 Flow map showing the movement of local trains between selected stations in West Bengal

Table 4.14 Worksheet for computing flow of water in tributary rivers and main river 'I'

Rivers or segment of rivers	Water discharge in Cusec (Cubic feet per second)	Scale selected	Width of the flow line (cm)
Tributary river 'A'	5	(1 cm width to 50 cusec water discharge)	0.1
Tributary river 'B'	6		0.12
Tributary river 'C'	10		0.2
Tributary river 'D'	6		0.12
Tributary river 'E'	8		0.16
Tributary river 'F'	5		0.1
Tributary river 'G'	8		0.16
Tributary river 'H'	7		0.14
Main river segment from P to Q	11		0.22
Main river segment from Q to S	21		0.42
Main river segment from R to S	14		0.28
Main river segment from S to U	35		0.7
Main river segment from T to U	13		0.26
Main river segment from U to V	48		0.96
Main river segment from V to W (Main river 'I')	55		1.1

Table 4.14 represents the discharge of water (in Cusec) in tributary rivers and the main river (River 'I') and this water discharge is graphically depicted in Fig. 4.43 with the help of flow lines with varying thickness. The width of the flow lines continuously increases towards the downstream section of the main river and it is due to the increase of the volume of water flowing towards downstream of the main river (River 'I').

4.12.7.2 Advantages and Disadvantages of Flow Map

Following are the advantages and disadvantages of flow map:

Advantages:

(1) Flow maps play great role to determine important transport routes and main centres of convergence (nodes) of goods or traffic. Pressure on different transport routes can easily be identified in this technique.

(2)   The sphere of influence of the nodes can easily be determined with the help of flow maps. It is done by measuring the thickness of the flow lines radiating outwards from the nodes. Generally, the thickness of the flow lines increases after some distances as one move towards an important place, a place to which people and commodities tend to move. This leads to divide the concerned area into a number of nodal regions (having nodal centres and sphere of influence).

(3)   Flow maps can again be subdivided into different types based on the type of commodity and the quantity flowed or moved.

Disadvantages:

(1)   Unavailability of accurate and reliable data regarding flow of people, traffic, goods etc. is the main difficulty in drawing flow map.

(2)   Some goods may show less value for large volume whereas the others may show more value for less volume. So the total load or tonnage may be misleading.

(3)   We may depend on the freight charges but their distribution is unidentified, because the charges are generally paid at some particular stations of certain routes though the goods or commodities may move over a number of routes.

(4)   Flow or movement in the same direction may create the problem of overlapping of lines.

(5)   Converging point of the wide lines is difficult to show without overwhelming the map.

4.12.8   Diagrammatic Map

Representation of the statistical (or geographical) data (Tables 4.15 and 4.16) on the map by means of suitable diagrams is called *Diagrammatic map*. It is a method of depiction of some kind of data within the boundary of particular administrative unit. Due to the application of diagrammatic languages in mapping techniques, the

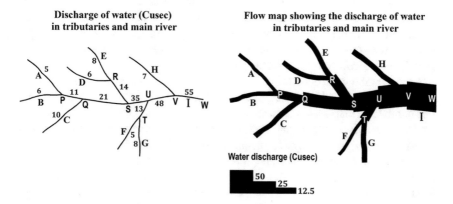

Fig. 4.43 Flow map showing the discharge of water in tributary rivers and main river (River 'I')

geographical precision of the map becomes discarded in favor of clarity. Major principles and procedures of diagrammatic mapping include—(a) the base of the diagrams like bar, rectangle, triangle etc. should be positioned at the exact centre of the region or at any point which may lie within the limits of the area concerned. (b) The centres of circular diagrams (like circles, spheres) must correspond to the exact place or to the areal centres of the administrative units. It needs the suitable selection of the scale of representation. (c) If the diagrams of adjacent administrative units appear to overlap, they should be drawn sequentially for places in the ascending order of size. The smaller ones must be clearly visible while the larger ones may be allowed to be partly consumed or eclipsed.

Diagrammatic maps are of different types (see Figs. 4.44 and 4.45):

(a)   Bar diagrammatic map
(b)   Circle diagrammatic map
(c)   Sphere diagrammatic map
(d)   Triangle diagrammatic map
(e)   Square diagrammatic map
(f)   Cube diagrammatic map etc.

4.13   Importance and Uses of Maps

Maps are one of the most *important tools of geographers*, researchers, cartographers, students and other concerned persons especially of geographers. Geographical study and analysis are much more dependent upon maps compared with other scientific disciplines. Geographers use different types of maps as '*shorthand scripts*' to analyze and understand the entire earth surface or a specific part of it. Maps help them to know about the whole world by showing the sizes and shapes of the continents, oceans, countries etc.; the locations and distributions of different physical and cultural features; the distances between places; areal characteristics of various features etc. They may be used as general reference to depict political boundaries, geomorphological features (say landforms), water bodies, vegetation-covered areas, soil types, the location of towns and cities etc. Maps are the guide and help to individuals in general and the government in particular.

Over the past few decades, the use of different types of maps has increased voluminously all over the world. It is mainly due to technological development that enables more numbers of map-making and their use than ever before. Greater human mobility, increase of analysis of geospatial relationships and more numbers of problems associated with various types of physical planning brought about by the intensive utilization of various natural resources like land, water, vegetation etc. are also important factors leading to continuous increase of use of maps. Presently, there has been an enormous increase in the number of different temporal maps, which are made visible only in soft copy on the monitor of a computer system but are not available in hard copy. Applying the same technological developments, large numbers of single or special purpose maps are being prepared, which are fully adjusted to the requirements of their users.

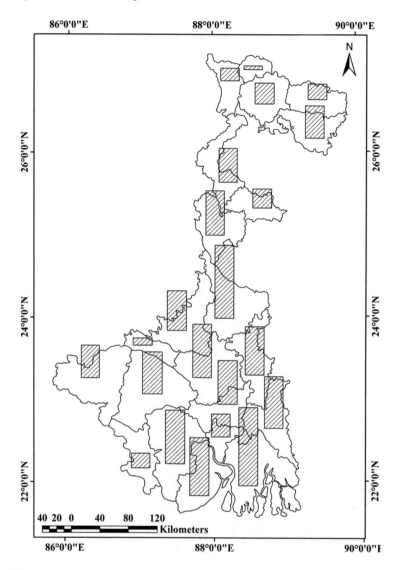

Fig. 4.44 Bar diagrammatic map showing the district-wise rural population in West Bengal (Census 2011)

Major uses and importance of maps are:

- Maps are the records of various facts and features of the earth that make a direct appeal to human mind. They may unfold the unknown and unseen lands and features in their true and original form.
- Complex relationships between the set of facts can easily be understood with the help of maps. For example, a population map with towns and cities of various sizes and kinds not only represent the factual data of population distribution in

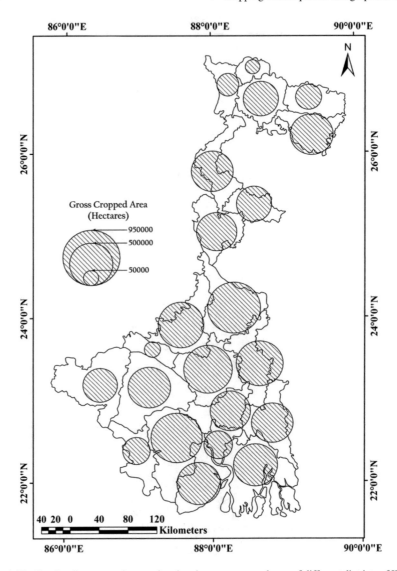

Fig. 4.45 Circular diagrammatic map showing the gross cropped area of different districts of West Bengal (2018–2019)

rural and urban areas but also explain the causal relationships that exist between them, i.e. it represents a concrete idea concerning other parts and people.

- Good maps provide us huge amount of information in its true perspectives. For example, topographical maps describe the regional geography of a region or country systematically and accurately.

- Navigational charts and maps are very important and popular for sea or air use.

Table 4.15 Calculations for bar diagrammatic map (District-wise rural population in West Bengal, census 2011)

Sl. Number	Name of the districts	Rural population	Scale selection	Height of the bar (cm)
1	Alipurduar[*]	1,183,704	1 cm height to 1,500,000 rural population	0.79
2	Bankura	3,296,901		2.20
3	Paschim Bardhaman	530,077		0.35
4	Purba Bardhaman	4,109,169		2.74
5	Birbhum	3,052,956		2.03
6	Cooch Behar	2,529,652		1.69
7	Darjeeling	923,410		0.62
8	Uttar Dinajpur	2,644,906		1.76
9	Dakshin Dinajpur	1,439,981		0.96
10	Hooghly	3,390,646		2.26
11	Howrah	1,775,885		1.18
12	Jalpaiguri	1,628,791		1.08
13	Jhargram[*]	1,096,996		0.73
14	Kolkata	0		0
15	Kalimpong[*]	195,450		0.13
16	Malda	3,447,185		2.30
17	Paschim Medinipur	4,093,775		2.73
18	Purba Medinipur	4,503,161		3.00
19	Murshidabad	5,703,115		3.80
20	Nadia	3,728,727		2.49
21	North 24 Parganas	4,277,619		2.85
22	South 24 Parganas	6,074,188		4.05
23	Purulia	2,556,801		1.70

[*] Was created after the 2011 census

Source Primary census abstract and district census report, 2011

- Importance of maps especially of topographical maps can never be neglected for military purposes. Maps become very helpful by indicating various routes and possible enemy positions during the time of operations and war.
- Weather reports, temperature, rainfall, wind direction, crops etc. can also be marked on the map.

Table 4.16 Calculations for circular diagrammatic map (Gross cropped area of different districts of West Bengal, 2018–2019)

Sl. Number	Name of the districts	Gross cropped area (GCA) in hectares	$r_i = \sqrt{\frac{T_i}{\Pi}}$	Scale selected	Radius of the circle (cm)
1	Alipurduar[*]	187,719	244.44	1 cm radius to 400 units	0.61
2	Bankura	541,194	415.05		1.04
3	Paschim Bardhaman	63,261	141.90		0.35
4	Purba Bardhaman	740,072	485.36		1.21
5	Birbhum	581,559	430.25		1.07
6	Cooch Behar	547,042	417.28		1.04
7	Darjeeling	153,106	220.76		0.55
8	Uttar Dinajpur	512,881	404.05		1.01
9	Dakshin Dinajpur	388,682	351.74		0.88
10	Hooghly	527,233	409.66		1.02
11	Howrah	165,003	299.18		0.75
12	Jalpaiguri	381,280	348.37		0.87
13	Jhargram[*]	221,060	265.26		0.66
14	Kolkata	0	0		0
15	Kalimpong[*]	56,696	134.34		0.33
16	Malda	467,038	385.57		0.96
17	Paschim Medinipur	832,316	514.72		1.29
18	Purba Medinipur	532,773	411.81		1.03
19	Murshidabad	937,356	546.23		1.36
20	Nadia	662,573	459.24		1.15
21	North 24 Parganas	500,385	399.09		1.00
22	South 24 Parganas	603,658	438.35		1.09
23	Purulia	369,801	343.09		0.86
For proportional scale	**Largest**	**950,000**	**549.90**		**1.37**
	Medium	**500,000**	**398.94**		**1.00**
	Smallest	**50,000**	**126.16**		**0.31**

[*] Was created after the 2011 census

Source   http://matirkatha.net/wp-content/uploads/2016/07/West-Bengal-District-wise-Cultivated-Area-Gross-cropped-are-cropping-intensity-2018-19.pdf

- Map helps us to calculate the distance between two places, mountains, rivers, cities, ports, railway stations etc. It can be used as a direction finder at minimum cost.
- It performs the role of a perfect guide in places which have never been visited before. Tourists and travellers don't need to depend on anybody to know the local directions if they have a map with them.
- It provides the information like the heights of different places or ups and downs found on the earth's surface. We also come to know the information about the presence of rivers, mountains, valleys or any other irregularities on the way of our journey for which we need to be prepared.
- It represents the boundaries of the lands and properties to define ownership. Government also uses the map to maintain the record of the owners.

Geographers make numerous measurements on map to determine directions, distances and area of various types of physical or cultural features.

4.13.1 Measurement of Direction

Direction of a point or place with respect to another point or place can easily be measured on a map. Direction on the map can be defined as the imaginary straight line indicating the angular position to a general base direction. The base direction is the line pointing towards the north having *zero direction*. A map always shows the true north direction, presented either by a north arrow or meridian. All other directions are easily determined in relation to this north direction. This true north direction becomes helpful to the map users for locating different aspects or features on the map with respect to each other. Four commonly used directions are North, South, East and West (also called the cardinal points) (Fig. 4.14, Sect. 4.8.9). Several intermediate directions may be present between these four main cardinal points.

4.13.1.1 Steps of Measurement of Direction

Following steps should be followed for the measurement of direction of points on map:

- Locate the points (M, N, O & P in Fig. 4.46) on map and connect them by straight lines.
- Draw north lines through the concerned points from which we have to measure the direction.
- Place the protector keeping 0° against the north line and the centre on the point from which we need to take the angular measurement.
- Take horizontal angular measurement from the north line to the connecting line to the right (clockwise).

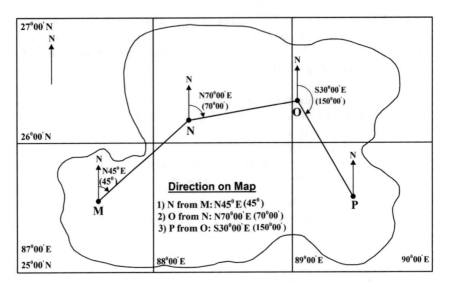

Fig. 4.46 Measurement of direction on map

- *Angular measurement* or direction may be expressed either in whole circle system as whole circle bearing ($360° > \theta > 0°$) or in quadrantal system as reduced bearing ($90° > \theta > 0°$), θ being the angular value.
- If *magnetic declination* is known to us then magnetic north direction can easily be drawn on the map by using the following equation:

$$\text{True bearing (Azimuth)} = \text{Magnetic bearing} \pm \text{Magnetic declination}$$

('+' is used in case of East magnetic declination and '−' is used in case of West magnetic declination) (Fig. 4.19 in Sect 4.8.12).

In Fig. 4.46, the location of four points (M, N, O and P) has been fixed and connected by MN, NO and OP lines. The directions of N, O and P points are measured from the north line using protector. Direction of N from M is N 45°00′ E (45°00′), O from N is N 70°00′ E (70°00′) and P from O is S 30°00′ E (150°00′).

4.13.2 Measurement of Distance

All types of linear features represented on the maps can be divided into two main categories, i.e. *straight line features* (like roads, railways, canals etc.) and *curved line or erratic or zigzag features* (like coastlines, rivers, streams etc.). The measurement of distance or length of any linear feature on the map is a function of the map scale. The degree of accuracy of measurement depends on the cartographic accuracy as well as on the type of scale used (Sarkar 2015). Suitable separate techniques are used for the measurement of distance of various straight and curved features on map.

4.13.2.1 Measurement of Distance of Straight Features

The *measurement of distance* or length of various straight map features such as roads, railway lines, canals etc. is very easy and simple. The map distance can be measured directly with a pair of dividers or a ruler (scale) placed on the map feature to be measured and then the map distance is converted into the actual ground distance (real-world distance) using the map's scale (by multiplying map distance by the denominator of the map scale or the scale factor).

Theoretically, if the

Map distance = D cm and

Map scale = 1 cm to S cm or km, then the

Actual ground distance = (D × S) cm or km.

For example, if the map distance of a given feature is 5 cm (D) on a map having a scale of 1:50,000 (S). Thus, the actual ground distance (real-world distance) of the concerned feature is 5 × 50,000 cm, i.e. 2,50,000 cm or 2.5 km.

4.13.2.2 Measurement of Distance of Curved Features

Measurement of distance of curved or *zigzag map features* like the coastlines, rivers, streams etc. is little more complicated than the measurement of distance of straight features. Generally, three techniques are popularly used for the measurement of distance of various curved map features. These include

(1)   Use of Ruler or Divider
(2)   Use of toned Thread
(3)   Use of Opisometer.

Use of Ruler or Divider

Ruler or divider can be used for the measurement of distance of different curved features. Ruler is used for this task to draw number of straight-line segments on the concerned curved feature for the measurement of map distance and then the map distance is converted into actual ground distance using the map scale. The accuracy of this technique depends on the number of straight-line segments used to measure the map distance. More the straight-line segments used more the degree of accuracy of the measurement (Fig. 4.47).

In case of using divider, at first, its legs are opened to cover a small convenient distance (length) say '*d*' cm. One leg of the divider is placed at the starting point of the curved feature and step off the feature from the starting point to the ending point. The number of steps (say N) required to reach the end point of the feature is counted properly. The map distance of the concerned feature is then calculated by multiplying the number of steps counted (N) by the small convenient distance (d cm in this case) [map distance = $N \times d$ cm]. This map distance is then converted into actual ground distance following the map scale (discussed in Sect. 4.13.2.1).

Use of Toned Thread

The use of *toned thread* is another important method to measure the distance of different curved line features on map. Steps of measurement of distance of curved features in this technique include:

- A knot is tied at one end of the thread as a reference point.
- The knot is then placed at the starting point of the curved feature and carrying the thread along the feature up to the end point so that it coincides exactly with the feature (Fig. 4.48).
- A mark is made on the thread when it reaches the other end of the feature.
- The thread is then removed from the feature, stretched and its map distance (length) is measured using ruler.
- Then the map distance of the thread is converted into actual ground distance of the feature using the given map scale (discussed in Sect. 4.13.2.1).

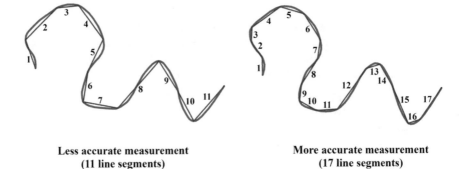

| Less accurate measurement | More accurate measurement |
| (11 line segments) | (17 line segments) |

Fig. 4.47 Measurement of distance of curved features on map using straight-line segments

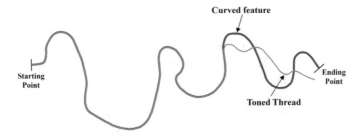

Fig. 4.48 Measurement of distance of curved features on map using toned thread

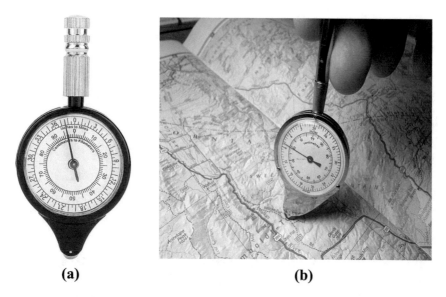

(a)   **(b)**

Fig. 4.49 An Opisometer (**a**) and the technique of measurement of distance of curved features on map using Opisometer (**b**)

Use of Opisometer

The map distance of *curvilinear features* can also be measured by using a mechanical device called an *opisometer*. It is also called a curvimeter or meilograph or map measurer. This device uses a small toothed wheel fitted with a recording dial. This toothed wheel is placed in contact with the curved feature to be measured and moves along its route, and the map distance is recorded directly in the dial (Fig. 4.49). The recorded distance is measured either in centimetres or inches. This map distance of the curved feature is converted into actual ground distance following the map scale (discussed in Sect. 4.13.2.1).

4.13.3   Measurement of Area

The *measurement of area* of different geographic units or features on the map with fair degree of accuracy is very important during a course of cartographic investigations and exercises by the map users especially by the students and scholars of geography. When administrative divisions (like country, province, state, districts, block etc.) are used as a basis of computations, the areas can exactly be measured from census volume, cadastral survey records or from large-scale maps on which the concerned areas are printed (Monkhouse and Wilkinson 1958). But when the map users deal with various non-administrative units like different types of landforms, forest-covered

Table 4.17 Methods of measurement of area on map

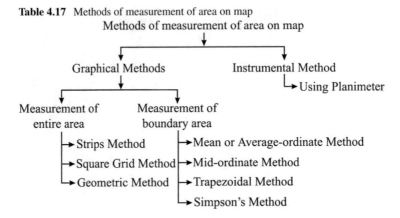

regions, erosion-prone regions, flood-prone regions etc., it is necessary to compute the area of these units because, in most of the cases, their area is not readily available to them. For example, in order to draw a hypsometric curve (also called hypsographic curve), the measurement of area within specific contours is necessary. Different graphical (including mathematical calculations) and instrumental methods (Table 4.17) are popularly used with greater or less degree of accuracy for the measurement of area of different features or units on map.

4.13.3.1 Graphical Methods

More often, the students and scholars of geography are associated with *irregular and asymmetric figures* of various shapes and sizes on which the formulae of regular geometric shapes cannot directly be applied for the measurement of the map area and, thus, there is the need for graphical approximation for their measurement.

Measurement of Entire Area

The area of the whole map can be measured using three important methods.

Strips method

This method is very fast but the degree of accuracy is comparatively less. Steps of measurement of area in this method include:

Step-1:   Series of parallel lines (called *strips*) of a fixed unit distance (called strip width, w) are drawn either upon the face of the map need to measure or on the tracing paper placed on the map. Width of the strip (w) is selected to

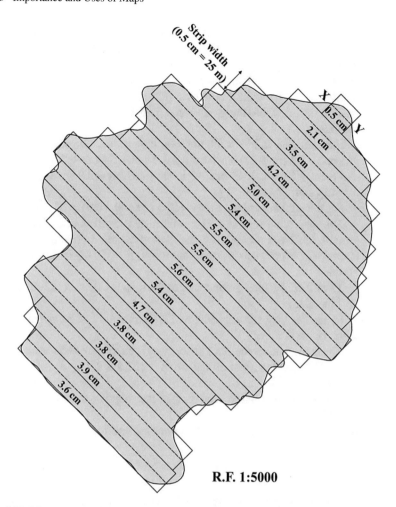

Fig. 4.50 Measurement of area on map by Strips method

represent a specific number of units (metres, kilometres etc.) following the scale of the map. For example, on a map with R.F. 1:50.00 (1 cm on map is equivalent to 50 m on ground), the strip width (w) of 0.5 cm represents 25 m on ground (Fig. 4.50). The smaller the width of the strip, the more precise will be the measurement.

Step-2: Strips are then converted into rectangles by drawing vertical lines at each end of every strip. The vertical lines should be drawn across each portion of the boundary of the map in such a way that the area to be included in the strip is as equal as possible to the area to be excluded from the strip (thus called 'give and take lines') (Fig. 4.50).

Step-3:   The length of each strip should be measured separately along a central line (like XY in the first strip) (Fig. 4.50), and the lengths of all the strips are then summed up to obtain the total strip length.

Step-4:   To obtain the area of the map, the total length of all the strips is then multiplied by the width of the strip (0.5 cm in Fig. 4.50). Then the map area is converted into the ground area (real-world area) using the scale of the map.

   For example, in Fig. 4.50, the total length of all the strips is 62.5 cm (0.5 cm + 2.1 cm + 3.5 cm + 4.2 cm + 5 cm + 5.4 cm + 5.5 cm + 5.5 cm + 5.6 cm + 5.4 cm + 4.7 cm + 3.8 cm + 3.8 cm + 3.9 cm + 3.6 cm) and the strip width (w) is 0.5 cm.

   So, the total area of the map is (62.5 cm × 0.5 cm) = 31.25 sq. cm.

   The map scale represents 1 cm on map is equivalent to 5,000 cm on ground, i.e. 1 cm on map is equivalent to 50 m on ground or, 1 sq. cm (1 cm × 1 cm) on map is equivalent to (50 m × 50 m) = 2,500 sq. m on ground.

   Therefore, the total ground area of the map is (31.25 × 2,500 sq. m) = 78,125 sq. m.

   In another way, the total length of all the strips is 62.5 cm and the strip width (w) is 0.5 cm.

   The map scale represents 1 cm on map is equivalent to 50 m on ground.

   So, the total length of all the strips is (62.5 × 50 m) = 3,125 m, and the strip width is 25 m on ground.

   Therefore, the total ground area of the map is (3,125 m × 25 m) = 78,125 sq. m.

Step-5:   Repeat the same procedure at least once for verification of the result.

Square grid method

Steps for the measurement of area on map using *grid square method* are as follows:

Step-1:   The area, need to measure should be traced on to a millimetre graph paper or a transparent millimetre graph paper should be superimposed on to the area need to measure. Operational Square of 2 mm × 2 mm (4 sq. mm) (Fig. 4.51) or 1 mm × 1 mm (1 sq. mm) dimension is then considered as the smallest unit of measurement of area. The smaller the unit square on the grid, the more accurate will be the measurement. But it should be noted that the minimum size of the square unit should be 1 mm × 1 mm = 1 sq. mm. If we consider 2 mm × 2 mm dimension as operational square, then the number of square in 1 sq. cm graph is 5 × 5 = 25 (Fig. 4.51).

Step-2:   Count the number of full squares of 2 mm × 2 mm dimension located in the area need to measure. Towards the centre of the concerned area, it is possible to count larger squares made. For example, in a 1 sq. cm larger square, the number of operational square of 2 mm × 2 mm dimension is 5 × 5 = 25 (Fig. 4.51). Similarly, in a 4 sq. cm larger square, the number

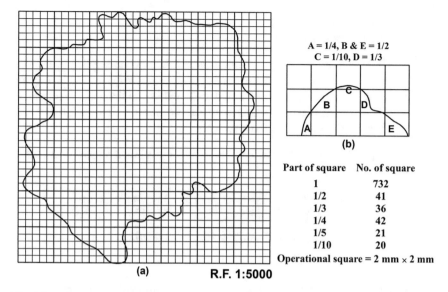

A = 1/4, B & E = 1/2
C = 1/10, D = 1/3

(b)

Part of square	No. of square
1	732
1/2	41
1/3	36
1/4	42
1/5	21
1/10	20

Operational square = 2 mm × 2 mm

(a)   R.F. 1:5000

Fig. 4.51 Measurement of area on map by square grid method

            of operational square of 2 mm × 2 mm dimension is $10 \times 10 = 100$. This makes our calculation less time consuming and easier.

Step-3:   Towards the edge or boundary, if more than one-half $(\frac{1}{2})$ portions of any square is located within the boundary of the concerned map area then it will be considered as a full square and the remaining edge squares will be ignored. The number of edge full squares is then added with the number of full squares calculated in step 2 to obtain the total number of full squares (say 'T'). The whole process should be repeated atleast once to verify the accuracy of the result.

Step-4:   The accuracy of the measurement can be improved by more precise measurement of the squares around the edge of the map, which are crossed by the map boundary. Edge squares can be measured in fractional (decimal) part of the whole square such as $\frac{1}{10}$ (0.1), $\frac{1}{5}$ (0.2), $\frac{1}{4}$ (0.25), $\frac{1}{3}$ (0.33), $\frac{1}{2}$ (0.5) etc. These fractional (decimal) parts of the squares are then converted in to full squares accordingly. Total number of full operational square (T) is obtained using the following formula (Sarkar 2015):

$$T = (\theta + \frac{\alpha}{10} + \frac{\beta}{5} + \frac{\phi}{4} + \frac{\gamma}{3} + \frac{\delta}{2}) \qquad (4.22)$$

where, θ is the number of full squares

α is the number of $\frac{1}{10\text{th}}$ squares.

β is the number of $\frac{1}{5\text{th}}$ squares.

ϕ is the number of $\frac{1}{4\text{th}}$ squares.

γ is the number of $\frac{1}{3rd}$ squares.

δ is the number of $\frac{1}{2nd}$ squares.

Step-5:   Total number of full *operational square* (T) is then multiplied by the area of each operational square to obtain the map area. If the operational square is of 2 mm × 2 mm dimension, then T will be multiplied by 4 sq. mm to obtain the map area. Then the map area is converted into the ground area (real-world area) using the map scale.

For example, in Fig. 4.51, total number of full operational square of 2 mm × 2 mm dimension,

$$T = \left(732 + \frac{20}{10} + \frac{21}{5} + \frac{42}{4} + \frac{36}{3} + \frac{41}{2}\right)$$
$$= (732 + 2 + 4.2 + 10.5 + 12 + 20.5)$$
$$= 781.2$$

The area of one operational square (2 mm × 2 mm dimension) is 4 sq. mm.
Therefore, the total area of the map is (781.2 × 4 sq. mm) = 3,124.8 sq. mm.
The map scale represents 1 cm (10 mm) on map is equivalent to 50 m on ground.
So, 2 mm on map = $\frac{50 \times 2}{10}$ = 10 m on ground.
Or, 4 sq. mm (2 mm × 2 mm) = $(10\,\text{m})^2$ = 100 sq. m.
Or, 3,124.8 sq. mm = $\frac{100 \times 3124.8}{4}$ sq. m = 78, 120 sq. m.
Thus, the ground area is 78, 120 sq. m.
In another way,
T = 781.2
The map scale represents 1 cm (10 mm) on map is equivalent to 50 m on ground.
So, 2 mm on map = $\frac{50 \times 2}{10}$ = 10 m on ground.
Map area of one operational square is = $(2\,\text{mm})^2$ = 4 sq. mm.
Or, ground area of one operational square is = $(10\,\text{m})^2$ = 100 sq. m.
Therefore, the total ground area is = 781.2 × 100 sq. m = 78, 120 sq. m.

Geometric Method

If the outline of the map unit is relatively simple then it is divided into different *regular geometrical shapes* like triangle, square, rectangle, trapezium etc. (Fig. 4.52) for the measurement of its area. A quadrilateral (ABCD in Fig. 4.52a) may be sub-divided into two triangles by joining two opposite angles. A polygon having more than four sides can be sub-divided into several triangles either by radiating from a central station (O) (Fig. 4.52b) or by radiating from a lateral station (P) (Fig. 4.52c). The *polygon* can also be sub-divided into several triangles and trapeziums using a baseline (OR) (Fig. 4.52d). The area of all the constituent geometric shapes is then added to obtain the total concerned area.

The *area of triangles* can be worked out by either of the formulae given below:

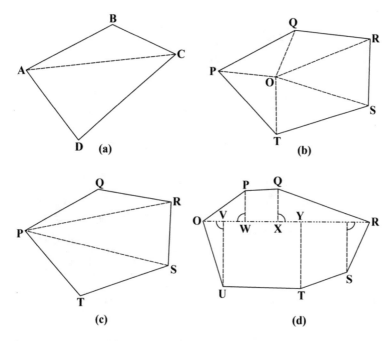

Fig. 4.52 Measurement of area by dividing into regular geometric shapes

(a) Area $= \sqrt{s\,(s-a)(s-b)(s-c)}$ (Any type of triangle especially of Scalene triangle)

$$(4.23)$$

where, a, b and c are the three sides of the triangle; s (semi perimeter) $= \frac{a+b+c}{2}$

(b) Area $= \frac{1}{2} \times$ Base $\times$ Perpendicular Height (Right angle triangle)   (4.24)

(c) Area $= \frac{1}{2} \times$ bc $\times$ Sin BAC (Any type of triangle)   (4.25)

where, b and c are two sides of the triangle and BAC is the included angle between them.

The area of a trapezium can be computed using the following formula:

$$\text{Area} = \text{Height} \times \frac{(\text{Base 1} + \text{Base 2})}{2} \tag{4.26}$$

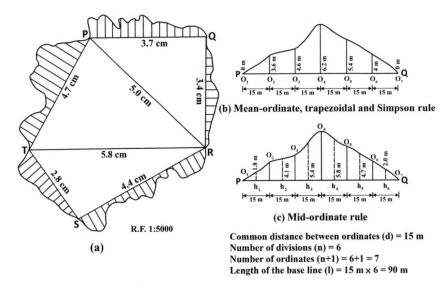

(c) Mid-ordinate rule

Common distance between ordinates (d) = 15 m
Number of divisions (n) = 6
Number of ordinates (n+1) = 6+1 = 7
Length of the base line (l) = 15 m × 6 = 90 m

(a)

Fig. 4.53 Measurement of area having irregular boundary using geometric method

where, base 1 is parallel to base 2 (for example, UV and TY in trapezium TUVY in Fig. 4.52d), and height is the perpendicular distance between base 1 and base 2 (VY in Fig. 4.52d).

Measurement of Boundary Area

For the measurement of map area having irregular outline, the entire area is divided into some regular geometric shapes, like triangles, occupying as much of the figure as possible (Fig. 4.53a) and their area are computed using either of the above mentioned formulae. But the problem arises regarding the computation of area of the *irregular portions* along the margins of the map (Fig. 4.53a). This problem of computing the area of the marginal portions can be solved using four possible methods.

1.   Mean or average ordinate method
2.   Mid ordinate method
3.   Trapezoidal Method
4.   Simpson Method

Mean or Average-Ordinate Method

In this method, perpendiculars (offsets) or ordinates (say $O_1, O_2, O_3 \cdots\cdots O_n$) are drawn at equal distance (say 'd') from the bounding lines of the triangles to the margin of the area (Fig. 4.53b). Closer the ordinates or offsets more accurate is the result. The area of the concerned irregular portion is computed using the following formula:

$$\text{Area} = \frac{\text{Sum of ordinates}}{\text{Number of ordinates}} \times \text{length of base line} \qquad (4.27)$$

$$\text{i.e. Area} = \frac{(O_1 + O_2 + O_3 + \cdots\cdots\ldots + O_n)}{n+1} \times l \qquad (4.28)$$

where,

O_1, O_2, O_3, ..., O_n are the lengths of each ordinate
l is the length of the baseline (i.e. $\sum d$ or nd)
d is the distance of each division (common distance between ordinates)
n is the number of divisions
$n + 1$ is the number of ordinates

In Fig. 4.53b, distance of each division $d = 15$ m.

the number of divisions $(n) = 6$
the number of ordinates $(n + 1) = 6 + 1 = 7$
length of the baseline $(l) = nd = 15m \times 6 = 90$ m.

Therefore, the

$$\text{Area} = \frac{(0\,\text{m} + 3.60\,\text{m} + 4.60\,\text{m} + 6.20\,\text{m} + 5.40\,\text{m} + 4.00\,\text{m} + 0\,\text{m})}{6+1} \times 90\,\text{m}.$$

$$= \frac{23.8\text{m}}{7} \times 90\,\text{m}.$$

$$= \frac{2142\,\text{sq. m.}}{7}$$

$$= 306\,\text{sq. m.}$$

Mid-ordinate Method

In this method, like average ordinate rule, the figure is divided into some strips by drawing offsets or ordinates (say $O_1, O_2, O_3, \ldots O_n$) at equal distance (say 'd') from the bounding lines of the triangles to the margin of the figure (Fig. 4.53c). Additionally, each strip is then sub-divided by drawing mid-ordinates (h_1, h_2, h_3 etc.). The ordinates are measured at the mid-ordinates of each strip and the area of the concerned irregular portion is computed using the following formula:

$$\text{Area} = (h_1 \times d) + (h_2 \times d) + \cdots + (h_{n-1} \times d) \qquad (4.29)$$

$$= d(h_1 + h_2 + \cdots + h_{n-1}) \qquad (4.30)$$

where,

$h_1, h_2, \ldots h_n$ are the lengths of mid-ordinates.
d is the common distance between ordinates.

i.e. Area = Common distance × Sum of mid-ordinates.
Again,

$$h_1 = \frac{O_1 + O_2}{2}$$

$$h_2 = \frac{O_2 + O_3}{2}$$

$$h_3 = \frac{O_3 + O_4}{2}$$

$$\ldots \ldots \ldots \ldots$$

$$h_{n-1} = \frac{O_{n-1} + O_n}{2}$$

In Fig. 4.53c,

$$h_1 = \frac{0\,m + 3.60\,m}{2} = 1.8\,m$$

$$h_2 = \frac{3.60\,m + 4.60\,m}{2} = 4.1\,m$$

$$h_3 = \frac{4.60\,m + 6.20\,m}{2} = 5.4\,m$$

$$h_4 = \frac{6.20\,m + 5.40\,m}{2} = 5.8\,m$$

$$h_5 = \frac{5.40\,m + 4.0\,m}{2} = 4.7\,m$$

$$h_6 = \frac{4.0\,m + 0\,m}{2} = 2.0\,m$$

And distance of each division $(d) = 15$ m.
Therefore, the

$$\text{Area} = 15\,m(1.8\,m + 4.1\,m + 5.4\,m + 5.8\,m + 4.7\,m + 2.0\,m)$$
$$= 15\,m \times 23.8\,m$$
$$= 357\,\text{sq.m.}$$

Trapezoidal Method

Measurement of area following *trapezoidal method* is more accurate than the previous two methods. In this method, the area enclosed by the irregular boundary line and the baseline is divided into trapezoids. It assumes that the boundaries between two extremities of the ordinates are straight lines. Following formula is used for the computation of area in this method:

$$\text{Area of 1st Trapezium} = \frac{O_1 \times O_2}{2}d$$

$$\text{Area of 2nd Trapezium} = \frac{O_2 \times O_3}{2}d$$

$$\text{Area of 3rd Trapezium} = \frac{O_3 \times O_4}{2}d$$

$$\text{Area of (n} - 1)\text{th Trapezium} = \frac{O_{n-1} \times O_n}{2}d$$

Therefore,

$$\text{The total area} = \frac{d}{2}(O_1 + 2O_2 + 2O_3 + \ldots + 2O_{n-1} + O_n) \tag{4.31}$$

$$= \frac{d}{2}[O_1 + O_n + 2(O_2 + O_3 + \ldots + O_{n-1})] \tag{4.32}$$

i.e. Area $= \frac{\text{common distance}}{2}$ [1st ordinate + last ordinate + 2(sum of other ordinates)]
In Fig. 4.53b,

$$\text{Total Area} = \frac{15\,\text{m}}{2}[0\,\text{m} + 0\,\text{m} + 2(3.60\,\text{m} + 4.60\,\text{m} + 6.20\,\text{m} + 5.40\,\text{m} + 4.00\,\text{m})]$$

$$= \frac{15\,\text{m}}{2}[0\,\text{m} + 0\,\text{m} + (2 \times 23.8\,\text{m})]$$

$$= \frac{15\,\text{m}}{2} \times [47.6\,\text{m}]$$

$$= \frac{714\,\text{sq. m.}}{2}$$

$$= 357\,\text{sq. m.}$$

Simpson Method

Inspite of having some complex geometric principles and calculations, this rule gives more accurate result than previously discussed other three methods. This rule assumes that the short lengths of boundary between the ordinates form an arc of parabola and hence *Simpson's method* is also called as *parabolic method*. In this method, the bounding line is divided into an odd number of offsets or ordinates to obtain an even number of divisions or unit areas, and the area is computed using the following formula:

Fig. 4.54 Principles of measurement of area using Simpson's method

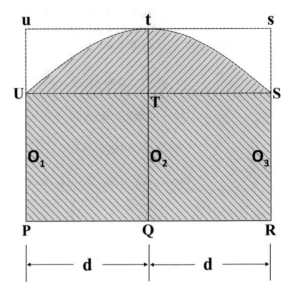

Total area $= \dfrac{d}{3}[O_1 + O_n + 4(O_2 + O_4 + \ldots) + 2(O_3 + O_5 + \ldots)]$

$$= \frac{\text{Common distance}}{3} \left[\begin{array}{l} \text{1st ordinate} + \text{Last ordinate} + 4(\text{Sum of even ordinates}) \\ +2(\text{Sum of remaining odd ordinates}) \end{array} \right]$$

$$(4.33)$$

Refer to Fig. 4.54,
Let

$O_1, O_2, O_3 =$ three successive ordinates
d is the common distance between ordinates.

According to Simpson's Rule:
The area between the first two divisions, Δ_1 (i.e. the area of PUtSR) = area of trapezium PUSR + area of portion UtSTU.
Here,

$$\text{Area of trapezium PUSR} = \frac{O_1 \times O_3}{2} \times 2d \qquad (4.34)$$

$$\text{Area of portion UtSTU} = \frac{2}{3} \times \text{Area of parallelogram UusS} = \frac{2}{3} \times Tt \times 2d$$

$$= \frac{2}{3} \left\{ O_2 - \frac{O_1 + O_3}{2} \right\} \times 2d$$

$$(4.35)$$

Therefore,

$$\text{Area of PUtSR} = \frac{O_1 \times O_3}{2} \times 2d + \frac{2}{3}\left\{O_2 - \frac{O_1 + O_3}{2}\right\} \times 2d \qquad (4.36)$$

$$= \frac{d}{3}(O_1 + 4O_2 + O_3) \qquad (4.37)$$

Similarly, the area between next two divisions (Δ_2)

$$= \frac{d}{3}(O_3 + 4O_4 + O_5) \qquad (4.38)$$

Thus, the total area of all the divisions can be computed as-

$$\text{Total area} = \frac{d}{3}[O_1 + O_n + 4(O_2 + O_4 + \dots) + 2(O_3 + O_5 + \dots)] \qquad (4.39)$$

In Fig. 4.53b,

$$\text{Total area} = \frac{15\,\text{m}}{3}[0\,\text{m} + 0\,\text{m} + 4(3.60\,\text{m} + 6.20\,\text{m} + 4.00\,\text{m}) + 2(4.60\,\text{m} + 5.40\,\text{m})]$$

$$= \frac{15\,\text{m}}{3}[0\,\text{m} + 0\,\text{m} + (4 \times 13.8\,\text{m}) + (2 \times 10\,\text{m})]$$

$$= \frac{15\,\text{m}}{3}[55.2\,\text{m} + 20\,\text{m}]$$

$$= \frac{15\,\text{m}}{3}[75.2\,\text{m}]$$

$$= \frac{1128\,\text{sq. m}}{3}$$

$$= 376\,\text{sq. m.}$$

4.13.3.2   Instrumental Method (Using Planimeter)

Graphical and geometric methods of computation of map area are only the approximate results and are laborious and time-consuming also. As a result, it has led to the development and use of mechanical or instrumental methods. The *planimeter* (Fig. 4.55), invented by Prof. J. Amsler, Swiss mathematician, is the only perfect instrument that can measure the map area by minimizing the error up to a limit of ±1%.

It is a small and handy but delicate instrument used for the determination of areas of all graphically represented regular and irregular shaped map units. Several models of the planimeter may be available—**(a) Hatchet planimeter**: It is the simplest and comprises a simple form of tracer bar. **(b) Polar or wheel planimeter**: It is more complex, delicate and equipped with automatic recording dial having either fixed or variable tracer arm. The planimeter having fixed tracer arm is called Fixed Arm

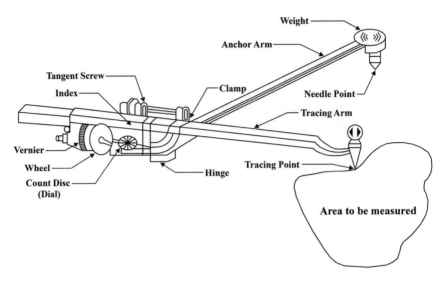

Fig. 4.55 Planimeter

Planimeter while the planimeter with variable tracer arm is known as Sliding Bar Planimeter (Singh and Singh 1991).

The main principle of each model of the planimeter is that a point is carefully and accurately traced along the outline (or perimeter) of the area to be measured in a clockwise direction. In case of a wheel planimeter, the total distance travelled while tracing the outline of the map is recorded on the recording dial, and this reading multiplied by the known constant for a particular instrument indicates the map area. Instrument with variable tracer arm will allow direct measurement of the map area in any unit (either in square centimetres or in square inches) while the instrument with fixed tracer arm measures the area on the map in square inches, and therefore the scale factor is used (Monkhouse and Wilkinson 1958). Two methods are there for the use of this instrument—(a) **The Interior Pole Method:** The needle pointed weight (fulcrum) is fixed inside the area to be measured and (b) **The Exterior Pole Method:** The needle pointed weight (fulcrum) is fixed outside the area to be measured. The latter method is more simple, reliable and accurate while the former is capable to cover a great range.

References

Burkard RK (1959) Geodesy for the Layman. U.S. Department of Commerce, National Oceanic and Atmospheric Administration, National Ocean Service

Kanetkar TP, Kulkarni SV (1984) Surveying and leveling Part I. Poona Vidhyarthi Griha Prakashan

Milliman JD, Meade RH (1983) Worldwide delivery of river sediments to the oceans. J Geol 91:1–21

Monkhouse FJ, Wilkinson HR (1958) Maps and diagrams: their compilation and construction. The Camelot Press Ltd., London and Southampton

Sarkar A (2015) Practical geography: a systematic approach. Orient Blackswan Private Limited, Hyderabad, Telangana, India. ISBN: 978-81-250-5903-5

Singh RL, Singh RPB (1991) Elements of practical geography. Kalyani Publishers

Wright JK (1936) A method of mapping densities of population with cape cod as an example. Geogr Rev 26(1):103–110

Web References

https://geographyandyou.com/the-geography-that-was-india/
http://physics.nmsu.edu/~jni/introgeophys/05_sea_surface_and_geoid/index.html
http://matirkatha.net/wp-content/uploads/2016/07/West-Bengal-District-wise-Cultivated-Area-Gross-cropped-are-cropping-intensity-2018-19.pdf
https://www.surveyofindia.gov.in/

Index

Printed in the United States
by Baker & Taylor Publisher Services